Spon's Estimating Costs Guide to Minor Landscaping, Gardening and External Works

Bryan Spain

Spon's Contractors' Handbooks

First published 2005 by Taylor & Francis
2 Park Square, Milton Park, Abingdon, Oxon, OX14 4RN

Simultaneously published in the USA and Canada by
Taylor & Francis
270 Madison Ave, New York, NY 10016

Taylor & Francis is an imprint of the Taylor & Francis Group

Publisher's Note
This book has been prepared from camera-ready copy supplied by the author.

Printed and bound in Great Britain by TJ International Ltd, Padstow, Cornwall

British Library Cataloguing in Publication Data
A catalogue record for this book is available from the British Library

Library of Congress Cataloguing in Publication Data
A catalog record for this book has been requested

ISBN 0-415-34410-7

Contents

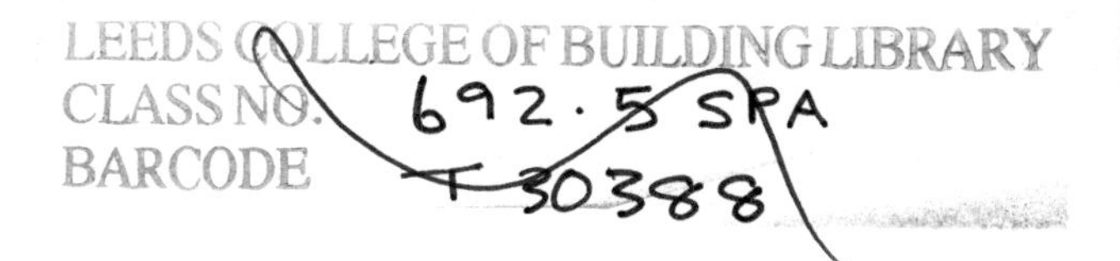

Spon's Estimating Costs Guide to Minor Landscaping and External Works

Other books by Bryan Spain

also available from Spon Press

Spon's Estimating Costs Guide to Roofing (2005 edition)	Pb: 0-415-34412-3
Spon's Estimating Costs Guide to Finishings: painting and decorating, plastering and tiling (2005 edition)	Pb: 0-415-34411-5
Spon's Estimating Costs Guide to Plumbing & Heating (2004 edition)	Pb: 0-415-31855-6
Spon's Estimating Costs Guide to Electrical Works (2004 edition)	Pb: 0-415-31853-X
Spon's Estimating Costs Guide to Minor Works, Alterations and Repairs to Fire, Flood, Gate and Theft Damage (2004 edition)	Pb: 0-415-31854-8
Spon's House Improvement Price Book: house extensions, storm damage work, alterations, loft conversions and insulation (2003 edition)	Pb: 0-415-30938-7
Spon's First Stage Estimating Price Book (2000 edition)	Pb: 0-415-23436-0
Spon's Construction Resource Handbook (1998 edition)	Hb: 0-419-23680-5

Preface

This edition of Spon's Estimating Costs Guide to Minor Landscaping and External Works is intended to provide accurate cost data for small landscaping and ground works contractors to prepare estimates and quotations more quickly and accurately.

This need for speed and accuracy is vital for all contractors operating in today's competitive construction market. Most contractors have the skills necessary to carry out the work together with the capacity for dealing with the setbacks that are part of the normal construction process. But they rarely have enough time to complete the many tasks that must be carried out in order to trade profitably. This book aims to help contractors by providing thousands of unit rates and, if used sensibly, can save them valuable time in the preparation of their bids..

I have received a great deal of support in the research necessary for this type of book and I am grateful to those individuals and firms who have provided the cost data and other information. In particular, I am indebted to Mark Loughrey of Youds, Ellison & Co., Chartered Accountants of Hoylake (tel: 0151-632 3298 or www.yesl.uk.com), who are specialists in advising small construction businesses. Their research for the information in the business section is based on tax legislation in force in December 2004.

Although every care has been taken in the preparation of the book, neither the Publisher nor I can accept any responsibility for the use of the information provided by any firm or individual. Finally, I would welcome any constructive criticism of the book's contents and suggestions that could be incorporated into future editions.

Bryan Spain
www.costofdiy.com
December 2004

Introduction

This edition of Spon's Estimating Costs Guide to Minor Landscaping and External Works follows the layout, style and contents of other books in this series. The contents of the book cover unit rates, project costs, repairs, tool and equipment hire, general advice on business matters and other information useful to those involved in the commissioning and construction of landscaping and external works. The unit rates section presents analytical rates for work up to about £50,000 in value and the business section covers advice on starting and running a business together with information on taxation and VAT matters.

Materials

In the domestic construction market, contractors are not usually able to purchase materials in large quantities and cannot benefit from the discounts available to larger contractors. An average of 10% to 15% discount has been allowed on normal trade prices.

Labour

The hourly labour rates for craftsmen and general operatives are based upon the current wage awards. These are set at:

Craftsman	£14.00
General operative	£11.00

These rates include provision for NIC Employers' contribution, CITB levy, insurances, public and annual holidays, severance pay and tool allowances where appropriate.

Headings

The following column headings have been used.

Unit	Labour	Hours £	Materials £	O & P £	Total £
m2	0.20	2.80	2.10	0.74	5.64

Unit

This column shows the unit of measurement for the item description:

nr	number
m	linear metre
m2	square metre
m3	cubic metre.

Labour

In the example shown, 0.20 represents the estimated time estimated to carry out one square metre of the described item, i.e. 0.20 hours.

Hours

The entry of £2.90 is calculated by multiplying the entry in the Labour column by the labour rate of £14.00.

Materials

This column displays the cost of the materials required to carry out one square metre of the described item, i.e. £2.10.

O & P (Overheads and profit)

This has been set at 15% and is deemed to cover head office and site overheads including:

- heating
- lighting
- rent
- rates
- telephones
- secretarial services
- insurances
- finance charges
- transport
- small tools
- ladders
- scaffolding etc.

Total

This is the total of the Hours, Materials and Overheads and Profit columns.

Contracting

Tradesmen and small contractors can act as main contractors (working for a client direct) or as a sub-contractor working for another contractor. Although a contract exists between a sub-contractor and a main contractor, there is no contractual link between a sub-contractor and an Employer.

In general terms this means that the sub-contractor cannot make any claims against the Employer direct and vice-versa. It also means that the sub-contractor should not accept any instructions from the Employer or his representative because this could be taken as establishing a privity of contract between the two parties,

A sub-contractor must be aware of his role in the programme because if he causes a contractor to overrun the completion date for the main contract he may become liable for the full amount of liquidated damages on the main contract plus the cost of damages that the contractor and other sub-contractors may have suffered.

A well-organised sub-contractor will keep a full set of daily site records, staffing levels, plant on site, weather charts and such like. It also cannot be over emphasised that any verbal instructions that the sub-contractor receives, should be confirmed immediately to the contractor in writing with the name of the person who issued them.

This procedure is extremely important because it may eventually save the sub-contractor considerable expense if someone tries to lay the blame for delays to the contract at his door. It is also important that instructions should only be taken from the contractor and he should be informed if another party attempts to do so.

Contractor's discount

Most sub-contracts allow for a discount to the contractor of 2½ % from the sub-contractor's account. This means that the sub-contractor must add this discount to his prices by adding 1/39th to his net rates.

Payment and retention

Payment is normally made on a monthly basis. The sub-contractor should submit his account to the contractor who then incorporates it into his own payment application and passes it on to the Architect or Employer's representative for certification of payment. When the sub-contractor receives his payment it will be reduced by 5% retention.

This money is held by the Employer and will be released in two parts. The first part, or moiety, is paid at the completion of work and, in the sub-contractor's case, this may be either when he has finished his work or when the contractor has completed the contract as a whole (known as practical completion) depending upon the contract conditions. The second part is released at the end of the defects liability period.

Defects liability period

This is the period of time (normally 12 months) during which the sub-contractor is contractually bound to return to the job to rectify any mistakes or bad pieces of workmanship. This could either be twelve months from when he completes his work or twelve months from when the main contract is completed depending upon the wording of the sub-contract.

Period for completion

Usually, a sub-contractor will be given a period of time in which he must complete the work and he must ensure that he has the capability to do the work within that period. Failure to meet the agreed completion date could have serious consequences.

Under certain circumstances, however, particularly with nominated sub-contracts, the sub-contractor may be requested to state the period of time he requires to do the work. If this is the case, then careful thought must be given to the time inserted. Too short a time may put him at financial risk but too long a time may prejudice the opportunity of winning the contract.

Damages for non-completion

A clause is usually inserted within each sub-contract stating that the sub-contractor is liable for the financial losses that contractor suffers due to the sub-contractor's non-completion of work on time. This will include the amount of liquidated and ascertained damages contained within the main contract, together with the contractor's own direct losses and the direct losses of his other sub-contractors. As can be seen, the potential cost to the sub-contractor can be large so he must take care to expedite the work with due diligence to avoid incurring these costs.

Variations

All sub-contracts contain a clause allowing the sub-contract work to be varied without invalidating it. The sub-contractor will normally be paid any additional cost he incurs in carrying out variations.

Insurances

The sub-contractor is responsible for insuring against injury to persons or property and against loss of plant and materials. These insurances could be taken out for each individual job, although it is more common to take out blanket policies based on the turnover the firm has achieved in the previous year.

Extensions of time

The sub-contractor will normally be entitled to a longer period of time to complete the work if he is delayed or interrupted by reasons beyond his control (known as an extension of time). Most sub-contracts list the reasons and in some cases the sub-contractor may also be entitled to additional monies as well as an extension of time.

Domestic sub-contracts

In domestic sub-contracts the contractor would obtain competitive quotations from various sub-contractors of his own choice and these may be based on a bill of quantities, specification and drawings, or schedules of work. Accompanying the enquiry should also be a form of sub-contract that the sub-contractor will be required to complete.

There are several points that may affect costs and which the sub-contractor should bear in mind. These are: -

1. Whether the rates and prices are to include for any contractor's discount (normally expressed as plus 1/39th to allow 2½%).

2. Whether the contractor is to supply any labour or plant to assist the sub-contractor in either carrying out any of the work or in off-loading materials.

3. What facilities (if any) the contractor will provide for the sub-contractor such as mess rooms, welfare facilities, office accommodation and storage facilities.

4. Whether the contractor is to dispose of the sub-contractor's rubbish.

Contracting

Often a sub-contractor will find himself working under a private contract, written or implied. This usually takes the form of working for a domestic householder or a small factory owner and the following procedures usually apply in this type of work.

Estimate

The initial approach would usually come from a purchaser, e.g. 'How much will it cost to have my garden landscaped?.' At this stage, he may only want an approximate cost in order to see if he can afford to have the work carried out, (as opposed to a quotation – see later). Therefore, a brief description of the work to be carried out together with an approximate price will suffice.

However, it should be made clear that the price is an estimate and does not constitute an offer that may be accepted by the purchaser. The estimate may be based on a telephone conversation only, e.g. 'It will cost about £4,000 to £ 5,000 to landscape your garden', or it could be based on a brief visit to the house. In either case, little time should be spent on an estimate and it is generally wise to express it as a price range.

Quotation

A quotation is generally seen as an offer to do the work for the price quoted, and could constitute a simple contract if accepted. It follows that some time and effort should be spent in compiling a quotation to save arguments at a later stage. One should always remember that the contractor is the expert and must use his expertise in order to guide the purchaser and should discuss the work with him in full. He should tell the purchaser exactly what he is getting for the price and also what he is not.

This may mean going in to some detail such as what will happen to the surplus excavated materials, how access will be gained, how long the job will take and similar items.

The contractor should also find out from the purchaser exactly what restrictions (if any) will be placed upon him. For instance, will the purchaser keep the drive clear of cars to allow a skip to be used and will the contractor only be allowed entry to the premises on certain days and/or at certain times? These factors, should be ascertained in advance, and the costs of complying with them should be made known to the purchaser who may decide to take steps to change the restrictions.

Once the contractor has considered all the relevant factors then the formal written quotation can be produced. It should state precisely what the purchaser is getting for his money, including when and how long the job will take and contain all the salient points of discussions that have taken place.

After a quotation has been submitted then all that needs to be done is for the purchaser to accept it. Although a verbal acceptance would constitute a binding agreement, it is always more satisfactory if the acceptance is made in writing.

Payments

There is much debate on how and when payments should be made in domestic situations. Ideally from a contractor's point of view to be paid in advance would be the most advantageous, but the chances of the purchaser wishing to do this are remote.

On the other hand, it may cause undue financial hardship to a recently self-employed contractor to have to buy all the materials himself and not get paid until all the work is completed. Whatever payment policy is adopted it must be agreed with the purchaser in advance and form part of the written quotation.

Possible alternatives are: -

1. Being paid when the work is complete. This is probably the best method from a public relations aspect and contractors who can complete a job in a few days should have no difficulty in adopting this policy.

2. Being paid before the work is done. This is only really feasible where the contractor concerned is of unquestionable reputation or is well known to the purchaser.

3. Being paid for materials as they are bought and delivered with the balance paid when the work is complete. This could be a practical solution for smaller contractors, but the purchaser will probably want proof of the material costs, so careful handling of invoices is necessary.

4. Some form of stage payments that usually take the form of agreed percentages of the quotation price or agreed parts of the quotation price paid after stages of the work have been carried out.

Pricing and variations

It is important that some method of recording, pricing and being paid for variations is agreed at the outset and this is particularly relevant when dealing with private clients. Unforeseen additions, more than any other item, are the main cause of disputes and are often avoidable.

The risk of this type of dispute can be reduced by ensuring that the original quotation is as detailed as possible. The detailed specification of the materials could be contained within the descriptions or done separately. A quotation broken down in this way is detailed enough to enable the purchaser to ascertain that he is not being overcharged for any variations that may occur and yet is not so detailed that the purchaser is going to question the price of every detail.

Also, if the purchaser should wish to change anything himself then there are no arguments on what was included in the original quotation.

If variations occur, it must be established who should pay for them. There are three main types of variations.

1. Those instructed by the purchaser.

2. Those that should have been included in the original quotation.

3. Those that are necessary due to events that could not have been foreseen.

The liabilities for 1 and 2 are relatively straightforward. If the purchaser says he wants a different paving flag to his original choice, then he must bear the additional cost. Conversely, if the contractor forgot to include the cost of the the sub-base in his quotation then it is only fair that he bears the cost.

Item 3 is more difficult. If it is the purchaser who is receiving the benefit of the variations and if they were not foreseeable, then it would be logical to assume that it is the purchaser who should bear the cost. An example would be where the excavation to a patio revealed old foundations underneath, the contractor would expect to be paid the extra cost for removing them.

Other instances may not be as clear cut as this example and it may become necessary to arrive at a cost-sharing arrangement if genuine doubt exists. Variations should preferably be agreed in advance before the work is carried out. They should be recorded and signed by both parties and, wherever possible, priced in detail and agreed.

Standard Method of Measurement/trades link

The contents of this book are presented under trade headings and the following table provides a link to the Standard Method of Measurement (SMM7).

External works

R12 Drainage below ground
Q40 Fencing
Q10 Concrete kerbs and edgings
Q20 Hardcore/granular sub-bases
Q25 Slab/brick pavings

Part One

UNIT RATES

Soft landscaping

Soil stabilisation

Excavation and filling

Seeding and turfing

Bare root trees

Container trees

Conifers

Shrubs

Climbers

Herbaceous plants

Hedging

Bedding plants

Maintenance

Sundries

	Unit	Labour	Hours £	Mat'ls £	O & P £	Total £

SOIL STABILISATION

Biogradable unseeded erosion control mats 2400mm wide fixed with pins to prepared ground

	Unit	Labour	Hours £	Mat'ls £	O & P £	Total £
Eromat Light	m2	0.05	0.70	2.36	0.46	3.52
Eromat Standard	m2	0.05	0.70	2.41	0.47	3.58
Eromat Coco	m2	0.05	0.70	2.44	0.47	3.61

Biogradable seeded erosion control mats 2400mm wide fixed with pins to prepared ground

	Unit	Labour	Hours £	Mat'ls £	O & P £	Total £
Covamat Standard	m2	0.05	0.70	2.69	0.51	3.90
Covamat Special	m2	0.05	0.70	2.87	0.54	4.11
Covamat Coco	m2	0.05	0.70	2.87	0.54	4.11

EXCAVATION AND FILLING

Excavation by hand

The following rates are based on excavating in firm ground. The following adjustments should be made for other conditions:

stiff clay + 50%
soft chalk + 100%

	Unit	Labour	Hours £	Mat'ls £	O & P £	Total £
Remove undergrowth and site vegetation	m2	0.15	2.10	-	0.32	2.42

	Unit	Labour	Hours £	Mat'ls £	O & P £	Total £
Cut down trees, grub up roots and remove						
girth, 600-1500mm	nr	20.00	280.00	-	42.00	322.00
girth, 1500-3000mm	nr	44.00	616.00	-	92.40	708.40
Cut down hedge, grub up roots and remove						
height, 1500mm	m	2.00	28.00	-	4.20	32.20
height, 3000mm	m	2.80	39.20	-	5.88	45.08
Excavate topsoil or turf and lay aside for re-use						
150mm thick	m2	0.35	4.90	-	0.74	5.64
200mm thick	m2	0.45	6.30	-	0.95	7.25
250mm thick	m2	0.60	8.40	-	1.26	9.66
Excavate to reduce levels depth not exceeding						
250mm thick	m3	2.20	30.80	-	4.62	35.42
500mm thick	m3	2.40	33.60	-	5.04	38.64
Excavate trenches depth not exceeding						
250mm thick	m3	2.40	33.60	-	5.04	38.64
500mm thick	m3	2.50	35.00	-	5.25	40.25
1000mm thick	m3	2.60	36.40	-	5.46	41.86
1500mm thick	m3	2.80	39.20	-	5.88	45.08

	Unit	Labour	Hours £	Mat'ls £	O & P £	Total £
Excavate pits depth not exceeding						
250mm thick	m3	2.60	36.40	-	5.46	41.86
500mm thick	m3	2.70	37.80	-	5.67	43.47
1000mm thick	m3	2.80	39.20	-	5.88	45.08
1500mm thick	m3	2.90	40.60	-	6.09	46.69
Extra for excavating through						
rock	m3	10.00	140.00	-	21.00	161.00
concrete	m3	8.00	112.00	-	16.80	128.80
brickwork	m3	6.00	84.00	-	12.60	96.60
Disposal by hand						
Load surplus excavated material into barrows, wheel and deposit in temporary spoil heaps, skip or lorry						
distance, 25m	m3	1.20	16.80	-	2.52	19.32
distance, 50m	m3	2.00	28.00	-	4.20	32.20
Load surplus excavated material into barrows, wheel and spread and level on site						
distance, 25m	m3	1.40	19.60	-	2.94	22.54
distance, 50m	m3	2.20	30.80	-	4.62	35.42

	Unit	Labour	Hours £	Mat'ls £	O & P £	Total £
Filling by hand						
Surplus excavated material deposited and compacted in layers						
over 250mm thick	m3	1.20	16.80	-	2.52	19.32
100mm thick	m2	0.20	2.80	-	0.42	3.22
150mm thick	m2	0.35	4.90	-	0.74	5.64
200mm thick	m2	0.50	7.00	-	1.05	8.05
Imported sand deposited and compacted in layers						
over 250mm thick	m3	1.20	16.80	31.24	7.21	55.25
100mm thick	m2	0.20	2.80	3.12	0.89	6.81
150mm thick	m2	0.35	4.90	4.68	1.44	11.02
200mm thick	m2	0.50	7.00	6.24	1.99	15.23
Imported hardcore deposited and compacted in layers						
over 250mm thick	m3	1.20	16.80	18.44	5.29	40.53
100mm thick	m2	0.20	2.80	1.84	0.70	5.34
150mm thick	m2	0.35	4.90	2.76	1.15	8.81
200mm thick	m2	0.50	7.00	3.68	1.60	12.28
Imported topsoil deposited and compacted in layers						
over 250mm thick	m3	1.20	16.80	14.00	4.62	35.42
100mm thick	m2	0.20	2.80	1.40	0.63	4.83
150mm thick	m2	0.35	4.90	2.10	1.05	8.05
200mm thick	m2	0.50	7.00	2.80	1.47	11.27

	Unit	Plant £	Mat'ls £	O & P £	Total £
Excavation by machine					
Where applicable the plant column includes the cost of the operator.					
The following rates are based on excavating in firm ground. The following adjustments should be made for other conditions:					
stiff clay + 50%					
soft chalk + 100%					
Remove undergrowth and site vegetation	m2	0.15	-	0.02	0.17
Cut down trees, grub up roots and remove					
girth, 600-1500mm	nr	184.00	-	27.60	211.60
girth, 1500-3000mm	nr	486.00	-	72.90	558.90
Cut down hedge, grub up roots and remove					
height, 1500mm	m	16.00	-	2.40	18.40
height, 3000mm	m	28.00	-	4.20	32.20
Excavate topsoil or turf and lay aside for re-use					
150mm thick	m2	0.30	-	0.05	0.35
200mm thick	m2	0.40	-	0.06	0.46
250mm thick	m2	0.50	-	0.08	0.58

	Unit	Plant £	Mat'ls £	O & P £	Total £
Excavate to reduce levels depth not exceeding					
250mm thick	m3	1.50	-	0.23	1.73
500mm thick	m3	1.40	-	0.21	1.61
Excavate trenches depth not exceeding					
250mm thick	m3	3.20	-	0.48	3.68
500mm thick	m3	3.10	-	0.47	3.57
1000mm thick	m3	3.00	-	0.45	3.45
1500mm thick	m3	2.90	-	0.44	3.34
Excavate pits depth not exceeding					
250mm thick	m3	3.30	-	0.50	3.80
500mm thick	m3	3.20	-	0.48	3.68
1000mm thick	m3	3.10	-	0.47	3.57
1500mm thick	m3	3.00	-	0.45	3.45
Extra for excavating through					
rock	m3	66.00	-	9.90	75.90
concrete	m3	56.00	-	8.40	64.40
brickwork	m3	44.00	-	6.60	50.60
Disposal by machine					
Load surplus excavated material into lorries and cart away to tip					
distance, 10km	m3	16.54	-	2.48	19.02
distance, 15km	m3	18.00	-	2.70	20.70

	Unit	Plant £	Mat'ls £	O & P £	Total £
Filling by machine					
Surplus excavated material deposited and compacted in layers					
over 250mm thick	m3	7.22	-	1.08	8.30
100mm thick	m2	1.10	-	0.17	1.27
150mm thick	m2	1.24	-	0.19	1.43
200mm thick	m2	1.38	-	0.21	1.59
Imported sand deposited and compacted in layers					
over 250mm thick	m3	7.22	31.24	5.77	44.23
100mm thick	m2	1.10	3.12	0.63	4.85
150mm thick	m2	1.24	4.68	0.89	6.81
200mm thick	m2	1.38	6.24	1.14	8.76
Imported hardcore deposited and compacted in layers					
over 250mm thick	m3	7.22	18.44	3.85	29.51
100mm thick	m2	1.10	1.84	0.44	3.38
150mm thick	m2	1.24	2.76	0.60	4.60
200mm thick	m2	1.38	3.68	0.76	5.82
Imported topsoil deposited and compacted in layers					
over 250mm thick	m3	7.22	14.00	3.18	24.40
100mm thick	m2	1.10	1.40	0.38	2.88
150mm thick	m2	1.24	2.10	0.50	3.84
200mm thick	m2	1.38	2.80	0.63	4.81

	Unit	Labour	Hours £	Mat'ls £	O & P £	Total £
SEEDING AND TURFING						
Imported topsoil deposited in spoil heaps	m3	-	-	14.00	2.10	16.10
Pre-seeding work by hand						
Lift topsoil from spoil heaps and spread and level in layers						
75mm thick	m2	0.03	0.42	-	0.06	0.48
100mm thick	m2	0.05	0.70	-	0.11	0.81
150mm thick	m2	0.07	0.98	-	0.15	1.13
Rake topsoil to a fine tilth	m2	0.04	0.56	-	0.08	0.64
Grass seeding by hand						
Sow grass seed on prepared ground						
PC £60.00 per 25kg						
10g per m2	m2	0.01	0.14	0.02	0.02	0.18
12g per m2	m2	0.01	0.14	0.02	0.02	0.18
14g per m2	m2	0.01	0.14	0.03	0.03	0.20
16g per m2	m2	0.01	0.14	0.03	0.03	0.20
18g per m2	m2	0.01	0.14	0.03	0.03	0.20
20g per m2	m2	0.01	0.14	0.04	0.03	0.21
22g per m2	m2	0.01	0.14	0.04	0.03	0.21
24g per m2	m2	0.01	0.14	0.04	0.03	0.21
26g per m2	m2	0.02	0.28	0.05	0.05	0.38
28g per m2	m2	0.02	0.28	0.05	0.05	0.38
30g per m2	m2	0.02	0.28	0.06	0.05	0.39

	Unit	Labour	Hours £	Mat'ls £	O & P £	Total £
32g per m2	m2	0.02	0.28	0.06	0.05	0.39
34g per m2	m2	0.02	0.28	0.06	0.05	0.39
36g per m2	m2	0.02	0.28	0.07	0.05	0.40
38g per m2	m2	0.02	0.28	0.07	0.05	0.40
40g per m2	m2	0.03	0.42	0.08	0.08	0.58
42g per m2	m2	0.03	0.42	0.08	0.08	0.58
44g per m2	m2	0.03	0.42	0.08	0.08	0.58
46g per m2	m2	0.03	0.42	0.09	0.08	0.59
48g per m2	m2	0.03	0.42	0.09	0.08	0.59
50g per m2	m2	0.03	0.42	0.10	0.08	0.60
PC £70.00 per 25kg						
10g per m2	m2	0.01	0.14	0.03	0.03	0.20
12g per m2	m2	0.01	0.14	0.03	0.03	0.20
14g per m2	m2	0.01	0.14	0.04	0.03	0.21
16g per m2	m2	0.01	0.14	0.04	0.03	0.21
18g per m2	m2	0.01	0.14	0.05	0.03	0.22
20g per m2	m2	0.01	0.14	0.06	0.03	0.23
22g per m2	m2	0.01	0.14	0.06	0.03	0.23
24g per m2	m2	0.01	0.14	0.07	0.03	0.24
26g per m2	m2	0.02	0.28	0.08	0.05	0.41
28g per m2	m2	0.02	0.28	0.08	0.05	0.41
30g per m2	m2	0.02	0.28	0.09	0.06	0.43
32g per m2	m2	0.02	0.28	0.09	0.06	0.43
34g per m2	m2	0.02	0.28	0.10	0.06	0.44
36g per m2	m2	0.02	0.28	0.11	0.06	0.45
38g per m2	m2	0.02	0.28	0.11	0.06	0.45
40g per m2	m2	0.03	0.42	0.12	0.08	0.62
42g per m2	m2	0.03	0.42	0.12	0.08	0.62
44g per m2	m2	0.03	0.42	0.13	0.08	0.63
46g per m2	m2	0.03	0.42	0.13	0.08	0.63
48g per m2	m2	0.03	0.42	0.14	0.08	0.64
50g per m2	m2	0.03	0.42	0.15	0.09	0.66

	Unit	Labour	Hours £	Mat'ls £	O & P £	Total £
Grass seeding by hand (cont'd)						
PC £80.00 per 25kg						
10g per m2	m2	0.01	0.14	0.03	0.03	0.20
12g per m2	m2	0.01	0.14	0.03	0.03	0.20
14g per m2	m2	0.01	0.14	0.04	0.03	0.21
16g per m2	m2	0.01	0.14	0.04	0.03	0.21
18g per m2	m2	0.01	0.14	0.05	0.03	0.22
20g per m2	m2	0.01	0.14	0.06	0.03	0.23
22g per m2	m2	0.01	0.14	0.06	0.03	0.23
24g per m2	m2	0.01	0.14	0.07	0.03	0.24
26g per m2	m2	0.02	0.28	0.08	0.05	0.41
28g per m2	m2	0.02	0.28	0.08	0.05	0.41
30g per m2	m2	0.02	0.28	0.09	0.06	0.43
32g per m2	m2	0.02	0.28	0.09	0.06	0.43
34g per m2	m2	0.02	0.28	0.10	0.06	0.44
36g per m2	m2	0.02	0.28	0.11	0.06	0.45
38g per m2	m2	0.02	0.28	0.11	0.06	0.45
40g per m2	m2	0.03	0.42	0.12	0.08	0.62
42g per m2	m2	0.03	0.42	0.12	0.08	0.62
44g per m2	m2	0.03	0.42	0.13	0.08	0.63
46g per m2	m2	0.03	0.42	0.13	0.08	0.63
48g per m2	m2	0.03	0.42	0.14	0.08	0.64
50g per m2	m2	0.03	0.42	0.15	0.09	0.66
PC £90.00 per 25kg						
10g per m2	m2	0.01	0.14	0.03	0.03	0.20
12g per m2	m2	0.01	0.14	0.03	0.03	0.20
14g per m2	m2	0.01	0.14	0.04	0.03	0.21
16g per m2	m2	0.01	0.14	0.04	0.03	0.21
18g per m2	m2	0.01	0.14	0.05	0.03	0.22
20g per m2	m2	0.01	0.14	0.06	0.03	0.23
22g per m2	m2	0.01	0.14	0.06	0.03	0.23
24g per m2	m2	0.01	0.14	0.07	0.03	0.24
26g per m2	m2	0.02	0.28	0.08	0.05	0.41

	Unit	Labour	Hours £	Mat'ls £	O & P £	Total £
28g per m2	m2	0.02	0.28	0.08	0.05	0.41
30g per m2	m2	0.02	0.28	0.09	0.06	0.43
32g per m2	m2	0.02	0.28	0.09	0.06	0.43
34g per m2	m2	0.02	0.28	0.10	0.06	0.44
36g per m2	m2	0.02	0.28	0.11	0.06	0.45
38g per m2	m2	0.02	0.28	0.11	0.06	0.45
40g per m2	m2	0.03	0.42	0.12	0.08	0.62
42g per m2	m2	0.03	0.42	0.12	0.08	0.62
44g per m2	m2	0.03	0.42	0.13	0.08	0.63
46g per m2	m2	0.03	0.42	0.13	0.08	0.63
48g per m2	m2	0.03	0.42	0.14	0.08	0.64
50g per m2	m2	0.03	0.42	0.15	0.09	0.66
PC £100.00 per 25kg						
10g per m2	m2	0.01	0.14	0.04	0.03	0.21
12g per m2	m2	0.01	0.14	0.05	0.03	0.22
14g per m2	m2	0.01	0.14	0.06	0.03	0.23
16g per m2	m2	0.01	0.14	0.06	0.03	0.23
18g per m2	m2	0.01	0.14	0.07	0.03	0.24
20g per m2	m2	0.01	0.14	0.08	0.03	0.25
22g per m2	m2	0.01	0.14	0.09	0.03	0.26
24g per m2	m2	0.01	0.14	0.10	0.04	0.28
26g per m2	m2	0.02	0.28	0.10	0.06	0.44
28g per m2	m2	0.02	0.28	0.11	0.06	0.45
30g per m2	m2	0.02	0.28	0.12	0.06	0.46
32g per m2	m2	0.02	0.28	0.13	0.06	0.47
34g per m2	m2	0.02	0.28	0.13	0.06	0.47
36g per m2	m2	0.02	0.28	0.14	0.06	0.48
38g per m2	m2	0.02	0.28	0.15	0.06	0.49
40g per m2	m2	0.03	0.42	0.16	0.09	0.67
42g per m2	m2	0.03	0.42	0.17	0.09	0.68
44g per m2	m2	0.03	0.42	0.17	0.09	0.68
46g per m2	m2	0.03	0.42	0.18	0.09	0.69
48g per m2	m2	0.03	0.42	0.19	0.09	0.70
50g per m2	m2	0.03	0.42	0.20	0.09	0.71

	Unit	Labour	Hours £	Mat'ls £	O & P £	Total £
Grass seeding by hand (cont'd)						
PC £110.00 per 25kg						
10g per m2	m2	0.01	0.14	0.04	0.03	0.21
12g per m2	m2	0.01	0.14	0.05	0.03	0.22
14g per m2	m2	0.01	0.14	0.06	0.03	0.23
16g per m2	m2	0.01	0.14	0.06	0.03	0.23
18g per m2	m2	0.01	0.14	0.07	0.03	0.24
20g per m2	m2	0.01	0.14	0.08	0.03	0.25
22g per m2	m2	0.01	0.14	0.09	0.03	0.26
24g per m2	m2	0.01	0.14	0.10	0.04	0.28
26g per m2	m2	0.02	0.28	0.10	0.06	0.44
28g per m2	m2	0.02	0.28	0.11	0.06	0.45
30g per m2	m2	0.02	0.28	0.12	0.06	0.46
32g per m2	m2	0.02	0.28	0.13	0.06	0.47
34g per m2	m2	0.02	0.28	0.13	0.06	0.47
36g per m2	m2	0.02	0.28	0.14	0.06	0.48
38g per m2	m2	0.02	0.28	0.15	0.06	0.49
40g per m2	m2	0.03	0.42	0.16	0.09	0.67
42g per m2	m2	0.03	0.42	0.17	0.09	0.68
44g per m2	m2	0.03	0.42	0.17	0.09	0.68
46g per m2	m2	0.03	0.42	0.18	0.09	0.69
48g per m2	m2	0.03	0.42	0.19	0.09	0.70
50g per m2	m2	0.03	0.42	0.20	0.09	0.71
PC £120.00 per 25kg						
10g per m2	m2	0.01	0.14	0.05	0.03	0.22
12g per m2	m2	0.01	0.14	0.06	0.03	0.23
14g per m2	m2	0.01	0.14	0.07	0.03	0.24
16g per m2	m2	0.01	0.14	0.08	0.03	0.25
18g per m2	m2	0.01	0.14	0.09	0.03	0.26
20g per m2	m2	0.01	0.14	0.10	0.04	0.28
22g per m2	m2	0.01	0.14	0.11	0.04	0.29
24g per m2	m2	0.01	0.14	0.12	0.04	0.30
26g per m2	m2	0.02	0.28	0.13	0.06	0.47

	Unit	Labour	Hours £	Mat'ls £	O & P £	Total £
28g per m2	m2	0.02	0.28	0.14	0.06	0.48
30g per m2	m2	0.02	0.28	0.15	0.06	0.49
32g per m2	m2	0.02	0.28	0.16	0.07	0.51
34g per m2	m2	0.02	0.28	0.17	0.07	0.52
36g per m2	m2	0.02	0.28	0.18	0.07	0.53
38g per m2	m2	0.02	0.28	0.19	0.07	0.54
40g per m2	m2	0.03	0.42	0.20	0.09	0.71
42g per m2	m2	0.03	0.42	0.21	0.09	0.72
44g per m2	m2	0.03	0.42	0.22	0.10	0.74
46g per m2	m2	0.03	0.42	0.23	0.10	0.75
48g per m2	m2	0.03	0.42	0.24	0.10	0.76
50g per m2	m2	0.03	0.42	0.25	0.10	0.77
PC £130.00 per 25kg						
10g per m2	m2	0.01	0.14	0.05	0.03	0.22
12g per m2	m2	0.01	0.14	0.06	0.03	0.23
14g per m2	m2	0.01	0.14	0.07	0.03	0.24
16g per m2	m2	0.01	0.14	0.08	0.03	0.25
18g per m2	m2	0.01	0.14	0.09	0.03	0.26
20g per m2	m2	0.01	0.14	0.10	0.04	0.28
22g per m2	m2	0.01	0.14	0.11	0.04	0.29
24g per m2	m2	0.01	0.14	0.12	0.04	0.30
26g per m2	m2	0.02	0.28	0.13	0.06	0.47
28g per m2	m2	0.02	0.28	0.14	0.06	0.48
30g per m2	m2	0.02	0.28	0.15	0.06	0.49
32g per m2	m2	0.02	0.28	0.16	0.07	0.51
34g per m2	m2	0.02	0.28	0.17	0.07	0.52
36g per m2	m2	0.02	0.28	0.18	0.07	0.53
38g per m2	m2	0.02	0.28	0.19	0.07	0.54
40g per m2	m2	0.03	0.42	0.20	0.09	0.71
42g per m2	m2	0.03	0.42	0.21	0.09	0.72
44g per m2	m2	0.03	0.42	0.22	0.10	0.74
46g per m2	m2	0.03	0.42	0.23	0.10	0.75
48g per m2	m2	0.03	0.42	0.24	0.10	0.76
50g per m2	m2	0.03	0.42	0.25	0.10	0.77

	Unit	Labour	Hours £	Mat'ls £	O & P £	Total £
Grass seeding by hand (cont'd)						
PC £140.00 per 25kg						
10g per m2	m2	0.01	0.14	0.05	0.03	0.22
12g per m2	m2	0.01	0.14	0.06	0.03	0.23
14g per m2	m2	0.01	0.14	0.07	0.03	0.24
16g per m2	m2	0.01	0.14	0.08	0.03	0.25
18g per m2	m2	0.01	0.14	0.09	0.03	0.26
20g per m2	m2	0.01	0.14	0.10	0.04	0.28
22g per m2	m2	0.01	0.14	0.11	0.04	0.29
24g per m2	m2	0.01	0.14	0.12	0.04	0.30
26g per m2	m2	0.02	0.28	0.13	0.06	0.47
28g per m2	m2	0.02	0.28	0.14	0.06	0.48
30g per m2	m2	0.02	0.28	0.15	0.06	0.49
32g per m2	m2	0.02	0.28	0.16	0.07	0.51
34g per m2	m2	0.02	0.28	0.17	0.07	0.52
36g per m2	m2	0.02	0.28	0.18	0.07	0.53
38g per m2	m2	0.02	0.28	0.19	0.07	0.54
40g per m2	m2	0.03	0.42	0.20	0.09	0.71
42g per m2	m2	0.03	0.42	0.21	0.09	0.72
44g per m2	m2	0.03	0.42	0.22	0.10	0.74
46g per m2	m2	0.03	0.42	0.23	0.10	0.75
48g per m2	m2	0.03	0.42	0.24	0.10	0.76
50g per m2	m2	0.03	0.42	0.25	0.10	0.77
PC £150.00 per 25kg						
10g per m2	m2	0.01	0.14	0.06	0.03	0.23
12g per m2	m2	0.01	0.14	0.07	0.03	0.24
14g per m2	m2	0.01	0.14	0.08	0.03	0.25
16g per m2	m2	0.01	0.14	0.10	0.04	0.28
18g per m2	m2	0.01	0.14	0.11	0.04	0.29
20g per m2	m2	0.01	0.14	0.12	0.04	0.30
22g per m2	m2	0.01	0.14	0.13	0.04	0.31
24g per m2	m2	0.01	0.14	0.14	0.04	0.32
26g per m2	m2	0.02	0.28	0.15	0.06	0.49

	Unit	Labour	Hours £	Mat'ls £	O & P £	Total £
28g per m2	m2	0.02	0.28	0.16	0.07	0.51
30g per m2	m2	0.02	0.28	0.18	0.07	0.53
32g per m2	m2	0.02	0.28	0.19	0.07	0.54
34g per m2	m2	0.02	0.28	0.20	0.07	0.55
36g per m2	m2	0.02	0.28	0.21	0.07	0.56
38g per m2	m2	0.02	0.28	0.22	0.08	0.58
40g per m2	m2	0.03	0.42	0.24	0.10	0.76
42g per m2	m2	0.03	0.42	0.25	0.10	0.77
44g per m2	m2	0.03	0.42	0.26	0.10	0.78
46g per m2	m2	0.03	0.42	0.27	0.10	0.79
48g per m2	m2	0.03	0.42	0.29	0.11	0.82
50g per m2	m2	0.03	0.42	0.30	0.11	0.83
PC £160.00 per 25kg						
10g per m2	m2	0.01	0.14	0.06	0.03	0.23
12g per m2	m2	0.01	0.14	0.07	0.03	0.24
14g per m2	m2	0.01	0.14	0.08	0.03	0.25
16g per m2	m2	0.01	0.14	0.10	0.04	0.28
18g per m2	m2	0.01	0.14	0.11	0.04	0.29
20g per m2	m2	0.01	0.14	0.12	0.04	0.30
22g per m2	m2	0.01	0.14	0.13	0.04	0.31
24g per m2	m2	0.01	0.14	0.14	0.04	0.32
26g per m2	m2	0.02	0.28	0.15	0.06	0.49
28g per m2	m2	0.02	0.28	0.16	0.07	0.51
30g per m2	m2	0.02	0.28	0.18	0.07	0.53
32g per m2	m2	0.02	0.28	0.19	0.07	0.54
34g per m2	m2	0.02	0.28	0.20	0.07	0.55
36g per m2	m2	0.02	0.28	0.21	0.07	0.56
38g per m2	m2	0.02	0.28	0.22	0.08	0.58
40g per m2	m2	0.03	0.42	0.24	0.10	0.76
42g per m2	m2	0.03	0.42	0.25	0.10	0.77
44g per m2	m2	0.03	0.42	0.26	0.10	0.78
46g per m2	m2	0.03	0.42	0.27	0.10	0.79
48g per m2	m2	0.03	0.42	0.29	0.11	0.82
50g per m2	m2	0.03	0.42	0.30	0.11	0.83

	Unit	Labour	Hours £	Mat'ls £	O & P £	Total £
Lawn treatment by hand by hand						
Apply lawn treatment						
bonemeal	m2	0.02	0.28	0.02	0.05	0.35
fish, blood and stone	m2	0.02	0.28	0.02	0.05	0.35
autumn/winter feed	m2	0.02	0.28	0.03	0.05	0.36
spring/summer feed	m2	0.02	0.28	0.03	0.05	0.36
weed and moss killer	m2	0.02	0.28	0.04	0.05	0.37
Turfing by hand						
Lay imported turf on prepared bed						
meadow turf	m2	0.08	1.12	1.12	0.34	2.58
sports outfields	m2	0.09	1.26	1.67	0.44	3.37
domestic lawns	m2	0.10	1.40	1.92	0.50	3.82

	Unit	Plant £	Mat'ls £	O & P £	Total £
Grass seeding by machine					
Where applicable the plant column includes the cost of the operator.					
Sow grass seed on prepared ground					
PC £60.00 per 25kg					
10g per m2	m2	0.01	0.02	0.00	0.03
12g per m2	m2	0.01	0.02	0.00	0.03
14g per m2	m2	0.01	0.03	0.01	0.05
16g per m2	m2	0.01	0.03	0.01	0.05
18g per m2	m2	0.01	0.03	0.01	0.05
20g per m2	m2	0.01	0.04	0.01	0.06
22g per m2	m2	0.01	0.04	0.01	0.06
24g per m2	m2	0.01	0.04	0.01	0.06
26g per m2	m2	0.01	0.05	0.01	0.07
28g per m2	m2	0.01	0.05	0.01	0.07
30g per m2	m2	0.01	0.06	0.01	0.08
32g per m2	m2	0.01	0.06	0.01	0.08
34g per m2	m2	0.01	0.06	0.01	0.08
36g per m2	m2	0.01	0.07	0.01	0.09
38g per m2	m2	0.01	0.07	0.01	0.09
40g per m2	m2	0.01	0.08	0.01	0.10
42g per m2	m2	0.01	0.08	0.01	0.10
44g per m2	m2	0.01	0.08	0.01	0.10
46g per m2	m2	0.01	0.09	0.02	0.12
48g per m2	m2	0.01	0.09	0.02	0.12
50g per m2	m2	0.01	0.10	0.02	0.13
PC £70.00 per 25kg					
10g per m2	m2	0.01	0.03	0.01	0.05
12g per m2	m2	0.01	0.03	0.01	0.05
14g per m2	m2	0.01	0.04	0.01	0.06

	Unit	Plant £	Mat'ls £	O & P £	Total £
Grass seeding by machine (cont'd)					
16g per m2	m2	0.01	0.04	0.01	0.06
18g per m2	m2	0.01	0.05	0.01	0.07
20g per m2	m2	0.01	0.06	0.01	0.08
22g per m2	m2	0.01	0.06	0.01	0.08
24g per m2	m2	0.01	0.07	0.01	0.09
26g per m2	m2	0.01	0.08	0.01	0.10
28g per m2	m2	0.01	0.08	0.01	0.10
30g per m2	m2	0.01	0.09	0.02	0.12
32g per m2	m2	0.01	0.09	0.02	0.12
34g per m2	m2	0.01	0.10	0.02	0.13
36g per m2	m2	0.01	0.11	0.02	0.14
38g per m2	m2	0.01	0.11	0.02	0.14
40g per m2	m2	0.01	0.12	0.02	0.15
42g per m2	m2	0.01	0.12	0.02	0.15
44g per m2	m2	0.01	0.13	0.02	0.16
46g per m2	m2	0.01	0.13	0.02	0.16
48g per m2	m2	0.01	0.14	0.02	0.17
50g per m2	m2	0.01	0.15	0.02	0.18
PC £80.00 per 25kg					
10g per m2	m2	0.01	0.03	0.01	0.05
12g per m2	m2	0.01	0.03	0.01	0.05
14g per m2	m2	0.01	0.04	0.01	0.06
16g per m2	m2	0.01	0.04	0.01	0.06
18g per m2	m2	0.01	0.05	0.01	0.07
20g per m2	m2	0.01	0.06	0.01	0.08
22g per m2	m2	0.01	0.06	0.01	0.08
24g per m2	m2	0.01	0.07	0.01	0.09
26g per m2	m2	0.01	0.08	0.01	0.10
28g per m2	m2	0.01	0.08	0.01	0.10
30g per m2	m2	0.01	0.09	0.02	0.12
32g per m2	m2	0.01	0.09	0.02	0.12
34g per m2	m2	0.01	0.10	0.02	0.13

	Unit	Plant £	Mat'ls £	O & P £	Total £
36g per m2	m2	0.01	0.11	0.02	0.14
38g per m2	m2	0.01	0.11	0.02	0.14
40g per m2	m2	0.01	0.12	0.02	0.15
42g per m2	m2	0.01	0.12	0.02	0.15
44g per m2	m2	0.01	0.13	0.02	0.16
46g per m2	m2	0.01	0.13	0.02	0.16
48g per m2	m2	0.01	0.14	0.02	0.17
50g per m2	m2	0.01	0.15	0.02	0.18
PC £90.00 per 25kg					
10g per m2	m2	0.01	0.03	0.01	0.05
12g per m2	m2	0.01	0.03	0.01	0.05
14g per m2	m2	0.01	0.04	0.01	0.06
16g per m2	m2	0.01	0.04	0.01	0.06
18g per m2	m2	0.01	0.05	0.01	0.07
20g per m2	m2	0.01	0.06	0.01	0.08
22g per m2	m2	0.01	0.06	0.01	0.08
24g per m2	m2	0.01	0.07	0.01	0.09
26g per m2	m2	0.01	0.08	0.01	0.10
28g per m2	m2	0.01	0.08	0.01	0.10
30g per m2	m2	0.01	0.09	0.02	0.12
32g per m2	m2	0.01	0.09	0.02	0.12
34g per m2	m2	0.01	0.10	0.02	0.13
36g per m2	m2	0.01	0.11	0.02	0.14
38g per m2	m2	0.01	0.11	0.02	0.14
40g per m2	m2	0.01	0.12	0.02	0.15
42g per m2	m2	0.01	0.12	0.02	0.15
44g per m2	m2	0.01	0.13	0.02	0.16
46g per m2	m2	0.01	0.13	0.02	0.16
48g per m2	m2	0.01	0.14	0.02	0.17
50g per m2	m2	0.01	0.15	0.02	0.18

	Unit	Plant £	Mat'ls £	O & P £	Total £
Grass seeding by machine (cont'd)					
PC £100.00 per 25kg					
10g per m2	m2	0.01	0.04	0.01	0.06
12g per m2	m2	0.01	0.05	0.01	0.07
14g per m2	m2	0.01	0.06	0.01	0.08
16g per m2	m2	0.01	0.06	0.01	0.08
18g per m2	m2	0.01	0.07	0.01	0.09
20g per m2	m2	0.01	0.08	0.01	0.10
22g per m2	m2	0.01	0.09	0.02	0.12
24g per m2	m2	0.01	0.10	0.02	0.13
26g per m2	m2	0.01	0.10	0.02	0.13
28g per m2	m2	0.01	0.11	0.02	0.14
30g per m2	m2	0.01	0.12	0.02	0.15
32g per m2	m2	0.01	0.13	0.02	0.16
34g per m2	m2	0.01	0.13	0.02	0.16
36g per m2	m2	0.01	0.14	0.02	0.17
38g per m2	m2	0.01	0.15	0.02	0.18
40g per m2	m2	0.01	0.16	0.03	0.20
42g per m2	m2	0.01	0.17	0.03	0.21
44g per m2	m2	0.01	0.17	0.03	0.21
46g per m2	m2	0.01	0.18	0.03	0.22
48g per m2	m2	0.01	0.19	0.03	0.23
50g per m2	m2	0.01	0.20	0.03	0.24
PC £110.00 per 25kg					
10g per m2	m2	0.01	0.04	0.01	0.06
12g per m2	m2	0.01	0.05	0.01	0.07
14g per m2	m2	0.01	0.06	0.01	0.08
16g per m2	m2	0.01	0.06	0.01	0.08
18g per m2	m2	0.01	0.07	0.01	0.09
20g per m2	m2	0.01	0.08	0.01	0.10
22g per m2	m2	0.01	0.09	0.02	0.12
24g per m2	m2	0.01	0.10	0.02	0.13
26g per m2	m2	0.01	0.10	0.02	0.13

	Unit	Plant £	Mat'ls £	O & P £	Total £
28g per m2	m2	0.01	0.11	0.02	0.14
30g per m2	m2	0.01	0.12	0.02	0.15
32g per m2	m2	0.01	0.13	0.02	0.16
34g per m2	m2	0.01	0.13	0.02	0.16
36g per m2	m2	0.01	0.14	0.02	0.17
38g per m2	m2	0.01	0.15	0.02	0.18
40g per m2	m2	0.01	0.16	0.03	0.20
42g per m2	m2	0.01	0.17	0.03	0.21
44g per m2	m2	0.01	0.17	0.03	0.21
46g per m2	m2	0.01	0.18	0.03	0.22
48g per m2	m2	0.01	0.19	0.03	0.23
50g per m2	m2	0.01	0.20	0.03	0.24
PC £120.00 per 25kg					
10g per m2	m2	0.01	0.05	0.01	0.07
12g per m2	m2	0.01	0.06	0.01	0.08
14g per m2	m2	0.01	0.07	0.01	0.09
16g per m2	m2	0.01	0.08	0.01	0.10
18g per m2	m2	0.01	0.09	0.02	0.12
20g per m2	m2	0.01	0.10	0.02	0.13
22g per m2	m2	0.01	0.11	0.02	0.14
24g per m2	m2	0.01	0.12	0.02	0.15
26g per m2	m2	0.01	0.13	0.02	0.16
28g per m2	m2	0.01	0.14	0.02	0.17
30g per m2	m2	0.01	0.15	0.02	0.18
32g per m2	m2	0.01	0.16	0.03	0.20
34g per m2	m2	0.01	0.17	0.03	0.21
36g per m2	m2	0.01	0.18	0.03	0.22
38g per m2	m2	0.01	0.19	0.03	0.23
40g per m2	m2	0.01	0.20	0.03	0.24
42g per m2	m2	0.01	0.21	0.03	0.25
44g per m2	m2	0.01	0.22	0.03	0.26
46g per m2	m2	0.01	0.23	0.04	0.28
48g per m2	m2	0.01	0.24	0.04	0.29
50g per m2	m2	0.01	0.25	0.04	0.30

	Unit	Plant £	Mat'ls £	O & P £	Total £
Grass seeding by machine (cont'd)					
PC £130.00 per 25kg					
10g per m2	m2	0.01	0.05	0.01	0.07
12g per m2	m2	0.01	0.06	0.01	0.08
14g per m2	m2	0.01	0.07	0.01	0.09
16g per m2	m2	0.01	0.08	0.01	0.10
18g per m2	m2	0.01	0.09	0.02	0.12
20g per m2	m2	0.01	0.10	0.02	0.13
22g per m2	m2	0.01	0.11	0.02	0.14
24g per m2	m2	0.01	0.12	0.02	0.15
26g per m2	m2	0.01	0.13	0.02	0.16
28g per m2	m2	0.01	0.14	0.02	0.17
30g per m2	m2	0.01	0.15	0.02	0.18
32g per m2	m2	0.01	0.16	0.03	0.20
34g per m2	m2	0.01	0.17	0.03	0.21
36g per m2	m2	0.01	0.18	0.03	0.22
38g per m2	m2	0.01	0.19	0.03	0.23
40g per m2	m2	0.01	0.20	0.03	0.24
42g per m2	m2	0.01	0.21	0.03	0.25
44g per m2	m2	0.01	0.22	0.03	0.26
46g per m2	m2	0.01	0.23	0.04	0.28
48g per m2	m2	0.01	0.24	0.04	0.29
50g per m2	m2	0.01	0.25	0.04	0.30
PC £140.00 per 25kg					
10g per m2	m2	0.01	0.05	0.01	0.07
12g per m2	m2	0.01	0.06	0.01	0.08
14g per m2	m2	0.01	0.07	0.01	0.09
16g per m2	m2	0.01	0.08	0.01	0.10
18g per m2	m2	0.01	0.09	0.02	0.12
20g per m2	m2	0.01	0.10	0.02	0.13
22g per m2	m2	0.01	0.11	0.02	0.14
24g per m2	m2	0.01	0.12	0.02	0.15
26g per m2	m2	0.01	0.13	0.02	0.16

	Unit	Plant £	Mat'ls £	O & P £	Total £
28g per m2	m2	0.01	0.14	0.02	0.17
30g per m2	m2	0.01	0.15	0.02	0.18
32g per m2	m2	0.01	0.16	0.03	0.20
34g per m2	m2	0.01	0.17	0.03	0.21
36g per m2	m2	0.01	0.18	0.03	0.22
38g per m2	m2	0.01	0.19	0.03	0.23
40g per m2	m2	0.01	0.20	0.03	0.24
42g per m2	m2	0.01	0.21	0.03	0.25
44g per m2	m2	0.01	0.22	0.03	0.26
46g per m2	m2	0.01	0.23	0.04	0.28
48g per m2	m2	0.01	0.24	0.04	0.29
50g per m2	m2	0.01	0.25	0.04	0.30
PC £150.00 per 25kg					
10g per m2	m2	0.01	0.06	0.01	0.08
12g per m2	m2	0.01	0.07	0.01	0.09
14g per m2	m2	0.01	0.08	0.01	0.10
16g per m2	m2	0.01	0.10	0.02	0.13
18g per m2	m2	0.01	0.11	0.02	0.14
20g per m2	m2	0.01	0.12	0.02	0.15
22g per m2	m2	0.01	0.13	0.02	0.16
24g per m2	m2	0.01	0.14	0.02	0.17
26g per m2	m2	0.01	0.15	0.02	0.18
28g per m2	m2	0.01	0.16	0.03	0.20
30g per m2	m2	0.01	0.18	0.03	0.22
32g per m2	m2	0.01	0.19	0.03	0.23
34g per m2	m2	0.01	0.20	0.03	0.24
36g per m2	m2	0.01	0.21	0.03	0.25
38g per m2	m2	0.01	0.22	0.03	0.26
40g per m2	m2	0.01	0.24	0.04	0.29
42g per m2	m2	0.01	0.25	0.04	0.30
44g per m2	m2	0.01	0.26	0.04	0.31
46g per m2	m2	0.01	0.27	0.04	0.32
48g per m2	m2	0.01	0.29	0.05	0.35
50g per m2	m2	0.01	0.30	0.05	0.36

	Unit	Plant £	Mat'ls £	O & P £	Total £
Grass seeding by machine (cont'd)					
PC £160.00 per 25kg					
10g per m2	m2	0.01	0.06	0.01	0.08
12g per m2	m2	0.01	0.07	0.01	0.09
14g per m2	m2	0.01	0.08	0.01	0.10
16g per m2	m2	0.01	0.10	0.02	0.13
18g per m2	m2	0.01	0.11	0.02	0.14
20g per m2	m2	0.01	0.12	0.02	0.15
22g per m2	m2	0.01	0.13	0.02	0.16
24g per m2	m2	0.01	0.14	0.02	0.17
26g per m2	m2	0.01	0.15	0.02	0.18
28g per m2	m2	0.01	0.16	0.03	0.20
30g per m2	m2	0.01	0.18	0.03	0.22
32g per m2	m2	0.01	0.19	0.03	0.23
34g per m2	m2	0.01	0.20	0.03	0.24
36g per m2	m2	0.01	0.21	0.03	0.25
38g per m2	m2	0.01	0.22	0.03	0.26
40g per m2	m2	0.01	0.24	0.04	0.29
42g per m2	m2	0.01	0.25	0.04	0.30
44g per m2	m2	0.01	0.26	0.04	0.31
46g per m2	m2	0.01	0.27	0.04	0.32
48g per m2	m2	0.01	0.29	0.05	0.35
50g per m2	m2	0.01	0.30	0.05	0.36
Lawn treatment by machine by hand					
Apply lawn treatment					
bonemeal	m2	0.01	0.02	0.00	0.03
fish, blood and stone	m2	0.01	0.02	0.00	0.03
autumn and winter feed	m2	0.01	0.03	0.01	0.05
spring amd summer feed	m2	0.01	0.03	0.01	0.05
weed and moss killer	m2	0.01	0.04	0.01	0.06

BARE ROOT TREES

The following descriptions refer to dimensions and sizes of trees as as set out below.

Whips: 125 to 175cm high
Feathered: 175 to 250cm high
Light standard: 6 to 8cm girth
Standard: 8 to 10cm girth
Selected standard: 10 to 12cm girth
Heavy standard: 12 to 14cm girth
Extra heavy standard: 14 to 16cm girth

	Unit	Labour	Hours £	Mat'ls £	O & P £	Total £
Excvavate tree pit, fork bottom, plant tree, backfill with excavated material including organic manure (30% of soil by volume), water and surround with peat						
Acer (Maple)						
campestre						
whip	nr	0.30	4.20	3.75	1.19	9.14
feathered	nr	0.40	5.60	8.05	2.05	15.70
light standard	nr	1.00	14.00	16.10	4.52	34.62
standard	nr	1.20	16.80	18.40	5.28	40.48
selected standard	nr	1.40	19.60	31.60	7.68	58.88
heavy standard	nr	1.60	22.40	51.75	11.12	85.27
extra heavy standard	nr	2.00	28.00	63.25	13.69	104.94
rubrum						
whip	nr	0.30	4.20	5.45	1.45	11.10
feathered	nr	0.40	5.60	11.50	2.57	19.67

	Unit	Labour	Hours £	Mat'ls £	O & P £	Total £
Tree planting (cont'd)						
light standard	nr	1.00	14.00	22.40	5.46	41.86
standard	nr	1.20	16.80	31.60	7.26	55.66
selected standard	nr	1.40	19.60	51.75	10.70	82.05
heavy standard	nr	1.60	22.40	63.25	12.85	98.50
extra heavy standard	nr	2.00	28.00	74.75	15.41	118.16
davidii						
light standard	nr	1.00	14.00	34.50	7.28	55.78
standard	nr	1.20	16.80	46.00	9.42	72.22
selected standard	nr	1.40	19.60	63.25	12.43	95.28
heavy standard	nr	1.60	22.40	92.00	17.16	131.56
extra heavy standard	nr	2.00	28.00	132.25	24.04	184.29
platanoides						
whip	nr	0.30	4.20	2.00	0.93	7.13
feathered	nr	0.40	5.60	5.20	1.62	12.42
light standard	nr	1.00	14.00	10.65	3.70	28.35
standard	nr	1.20	16.80	12.95	4.46	34.21
selected standard	nr	1.40	19.60	18.70	5.75	44.05
heavy standard	nr	1.60	22.40	27.60	7.50	57.50
extra heavy standard	nr	2.00	28.00	63.25	13.69	104.94
Aesculus (Chestnut)						
hippocastanum						
whip	nr	0.30	4.20	3.75	1.19	9.14
feathered	nr	0.40	5.60	8.05	2.05	15.70
light standard	nr	1.00	14.00	16.10	4.52	34.62
standard	nr	1.20	16.80	18.40	5.28	40.48
selected standard	nr	1.40	19.60	31.60	7.68	58.88
heavy standard	nr	1.60	22.40	51.75	11.12	85.27
extra heavy standard	nr	2.00	28.00	63.25	13.69	104.94

	Unit	Labour	Hours £	Mat'ls £	O & P £	Total £
h. baumanii						
light standard	nr	1.00	14.00	34.50	7.28	55.78
standard	nr	1.20	16.80	46.00	9.42	72.22
selected standard	nr	1.40	19.60	63.25	12.43	95.28
heavy standard	nr	1.60	22.40	92.00	17.16	131.56
extra heavy standard	nr	2.00	28.00	132.25	24.04	184.29
Alnus (Alder)						
glutinosa						
whip	nr	0.30	4.20	2.60	1.02	7.82
feathered	nr	0.40	5.60	6.30	1.79	13.69
light standard	nr	1.00	14.00	13.50	4.13	31.63
standard	nr	1.20	16.80	15.55	4.85	37.20
selected standard	nr	1.40	19.60	25.30	6.74	51.64
heavy standard	nr	1.60	22.40	46.00	10.26	78.66
extra heavy standard	nr	2.00	28.00	57.50	12.83	98.33
g. laciniata						
light standard	nr	1.00	14.00	34.50	7.28	55.78
standard	nr	1.20	16.80	46.00	9.42	72.22
selected standard	nr	1.40	19.60	63.25	12.43	95.28
heavy standard	nr	1.60	22.40	92.00	17.16	131.56
extra heavy standard	nr	2.00	28.00	132.25	24.04	184.29
Betula (Birch)						
albosinensis						
light standard	nr	1.00	14.00	34.50	7.28	55.78
standard	nr	1.20	16.80	46.00	9.42	72.22
selected standard	nr	1.40	19.60	63.25	12.43	95.28
heavy standard	nr	1.60	22.40	92.00	17.16	131.56
extra heavy standard	nr	2.00	28.00	132.25	24.04	184.29

	Unit	Labour	Hours £	Mat'ls £	O & P £	Total £
Tree planting (cont'd)						
ermanii						
light standard	nr	1.00	14.00	34.50	7.28	55.78
standard	nr	1.20	16.80	46.00	9.42	72.22
selected standard	nr	1.40	19.60	63.25	12.43	95.28
heavy standard	nr	1.60	22.40	92.00	17.16	131.56
extra heavy standard	nr	2.00	28.00	132.25	24.04	184.29
pendula						
whip	nr	0.30	4.20	2.00	0.93	7.13
feathered	nr	0.40	5.60	5.20	1.62	12.42
light standard	nr	1.00	14.00	10.65	3.70	28.35
standard	nr	1.20	16.80	12.95	4.46	34.21
selected standard	nr	1.40	19.60	18.70	5.75	44.05
heavy standard	nr	1.60	22.40	27.60	7.50	57.50
extra heavy standard	nr	2.00	28.00	63.25	13.69	104.94
p. youngii						
light standard	nr	1.00	14.00	34.50	7.28	55.78
standard	nr	1.20	16.80	46.00	9.42	72.22
selected standard	nr	1.40	19.60	63.25	12.43	95.28
heavy standard	nr	1.60	22.40	92.00	17.16	131.56
extra heavy standard	nr	2.00	28.00	132.25	24.04	184.29
Fagus (Beech)						
sylvatica						
whip	nr	0.30	4.20	3.75	1.19	9.14
feathered	nr	0.40	5.60	8.05	2.05	15.70
light standard	nr	1.00	14.00	16.10	4.52	34.62
standard	nr	1.20	16.80	18.40	5.28	40.48
selected standard	nr	1.40	19.60	31.60	7.68	58.88
heavy standard	nr	1.60	22.40	51.75	11.12	85.27
extra heavy standard	nr	2.00	28.00	63.25	13.69	104.94

	Unit	Labour	Hours £	Mat'ls £	O & P £	Total £
s. Dawyck purple						
light standard	nr	1.00	14.00	34.50	7.28	55.78
standard	nr	1.20	16.80	46.00	9.42	72.22
selected standard	nr	1.40	19.60	63.25	12.43	95.28
heavy standard	nr	1.60	22.40	92.00	17.16	131.56
extra heavy standard	nr	2.00	28.00	132.25	24.04	184.29
Fraxinus (Ash)						
angustifolia Raywood						
whip	nr	0.30	4.20	3.75	1.19	9.14
feathered	nr	0.40	5.60	8.05	2.05	15.70
light standard	nr	1.00	14.00	16.10	4.52	34.62
standard	nr	1.20	16.80	18.40	5.28	40.48
selected standard	nr	1.40	19.60	31.60	7.68	58.88
heavy standard	nr	1.60	22.40	51.75	11.12	85.27
extra heavy standard	nr	2.00	28.00	63.25	13.69	104.94
jaspidea						
whip	nr	0.30	4.20	5.45	1.45	11.10
feathered	nr	0.40	5.60	11.50	2.57	19.67
light standard	nr	1.00	14.00	22.40	5.46	41.86
standard	nr	1.20	16.80	31.60	7.26	55.66
selected standard	nr	1.40	19.60	51.75	10.70	82.05
heavy standard	nr	1.60	22.40	63.25	12.85	98.50
extra heavy standard	nr	2.00	28.00	74.75	15.41	118.16
pendula						
whip	nr	0.30	4.20	5.45	1.45	11.10
feathered	nr	0.40	5.60	11.50	2.57	19.67
light standard	nr	1.00	14.00	22.40	5.46	41.86
standard	nr	1.20	16.80	31.60	7.26	55.66
selected standard	nr	1.40	19.60	51.75	10.70	82.05
heavy standard	nr	1.60	22.40	63.25	12.85	98.50
extra heavy standard	nr	2.00	28.00	74.75	15.41	118.16

	Unit	Labour	Hours £	Mat'ls £	O & P £	Total £
Tree planting (cont'd)						
Prunus (Cherry)						
avium						
whip	nr	0.30	4.20	2.00	0.93	7.13
feathered	nr	0.40	5.60	5.20	1.62	12.42
light standard	nr	1.00	14.00	10.65	3.70	28.35
standard	nr	1.20	16.80	12.95	4.46	34.21
selected standard	nr	1.40	19.60	18.70	5.75	44.05
heavy standard	nr	1.60	22.40	27.60	7.50	57.50
extra heavy standard	nr	2.00	28.00	63.25	13.69	104.94
dulcis						
whip	nr	0.30	4.20	3.75	1.19	9.14
feathered	nr	0.40	5.60	8.05	2.05	15.70
light standard	nr	1.00	14.00	16.10	4.52	34.62
standard	nr	1.20	16.80	18.40	5.28	40.48
selected standard	nr	1.40	19.60	31.60	7.68	58.88
heavy standard	nr	1.60	22.40	51.75	11.12	85.27
extra heavy standard	nr	2.00	28.00	63.25	13.69	104.94
royal burgundy						
whip	nr	0.30	4.20	5.45	1.45	11.10
feathered	nr	0.40	5.60	11.50	2.57	19.67
light standard	nr	1.00	14.00	22.40	5.46	41.86
standard	nr	1.20	16.80	31.60	7.26	55.66
selected standard	nr	1.40	19.60	51.75	10.70	82.05
heavy standard	nr	1.60	22.40	63.25	12.85	98.50
extra heavy standard	nr	2.00	28.00	74.75	15.41	118.16
taihaku						
whip	nr	0.30	4.20	3.75	1.19	9.14
feathered	nr	0.40	5.60	8.05	2.05	15.70
light standard	nr	1.00	14.00	16.10	4.52	34.62
standard	nr	1.20	16.80	18.40	5.28	40.48

	Unit	Labour	Hours £	Mat'ls £	O & P £	Total £
selected standard	nr	1.40	19.60	31.60	7.68	58.88
heavy standard	nr	1.60	22.40	51.75	11.12	85.27
extra heavy standard	nr	2.00	28.00	63.25	13.69	104.94
Quercus (Oak)						
cerris						
whip	nr	0.30	4.20	5.45	1.45	11.10
feathered	nr	0.40	5.60	11.50	2.57	19.67
light standard	nr	1.00	14.00	22.40	5.46	41.86
standard	nr	1.20	16.80	31.60	7.26	55.66
selected standard	nr	1.40	19.60	51.75	10.70	82.05
heavy standard	nr	1.60	22.40	63.25	12.85	98.50
extra heavy standard	nr	2.00	28.00	74.75	15.41	118.16
robur						
whip	nr	0.30	4.20	2.60	1.02	7.82
feathered	nr	0.40	5.60	6.30	1.79	13.69
light standard	nr	1.00	14.00	13.50	4.13	31.63
standard	nr	1.20	16.80	15.55	4.85	37.20
selected standard	nr	1.40	19.60	25.30	6.74	51.64
heavy standard	nr	1.60	22.40	46.00	10.26	78.66
extra heavy standard	nr	2.00	28.00	57.50	12.83	98.33
rubra						
whip	nr	0.30	4.20	3.75	1.19	9.14
feathered	nr	0.40	5.60	8.05	2.05	15.70
light standard	nr	1.00	14.00	16.10	4.52	34.62
standard	nr	1.20	16.80	18.40	5.28	40.48
selected standard	nr	1.40	19.60	31.60	7.68	58.88
heavy standard	nr	1.60	22.40	51.75	11.12	85.27
extra heavy standard	nr	2.00	28.00	63.25	13.69	104.94

	Unit	Labour	Hours £	Mat'ls £	O & P £	Total £
Tree planting (cont'd)						
Sorbus (Whitebeam)						
aria						
whip	nr	0.30	4.20	2.60	1.02	7.82
feathered	nr	0.40	5.60	6.30	1.79	13.69
light standard	nr	1.00	14.00	13.50	4.13	31.63
standard	nr	1.20	16.80	15.55	4.85	37.20
selected standard	nr	1.40	19.60	25.30	6.74	51.64
heavy standard	nr	1.60	22.40	46.00	10.26	78.66
extra heavy standard	nr	2.00	28.00	57.50	12.83	98.33
aucuparia						
whip	nr	0.30	4.20	2.00	0.93	7.13
feathered	nr	0.40	5.60	5.20	1.62	12.42
light standard	nr	1.00	14.00	10.65	3.70	28.35
standard	nr	1.20	16.80	12.95	4.46	34.21
selected standard	nr	1.40	19.60	18.70	5.75	44.05
heavy standard	nr	1.60	22.40	27.60	7.50	57.50
extra heavy standard	nr	2.00	28.00	63.25	13.69	104.94
hupehensis						
whip	nr	0.30	4.20	3.75	1.19	9.14
feathered	nr	0.40	5.60	8.05	2.05	15.70
light standard	nr	1.00	14.00	16.10	4.52	34.62
standard	nr	1.20	16.80	18.40	5.28	40.48
selected standard	nr	1.40	19.60	31.60	7.68	58.88
heavy standard	nr	1.60	22.40	51.75	11.12	85.27
extra heavy standard	nr	2.00	28.00	63.25	13.69	104.94
sargentiana						
whip	nr	0.30	4.20	5.45	1.45	11.10
feathered	nr	0.40	5.60	11.50	2.57	19.67
light standard	nr	1.00	14.00	22.40	5.46	41.86
standard	nr	1.20	16.80	31.60	7.26	55.66

	Unit	Labour	Hours £	Mat'ls £	O & P £	Total £
selected standard	nr	1.40	19.60	51.75	10.70	82.05
heavy standard	nr	1.60	22.40	63.25	12.85	98.50
extra heavy standard	nr	2.00	28.00	74.75	15.41	118.16
torminalis						
whip	nr	0.30	4.20	5.45	1.45	11.10
feathered	nr	0.40	5.60	11.50	2.57	19.67
light standard	nr	1.00	14.00	22.40	5.46	41.86
standard	nr	1.20	16.80	31.60	7.26	55.66
selected standard	nr	1.40	19.60	51.75	10.70	82.05
heavy standard	nr	1.60	22.40	63.25	12.85	98.50
extra heavy standard	nr	2.00	28.00	74.75	15.41	118.16
Tilia (Lime)						
cordata						
whip	nr	0.30	4.20	2.60	1.02	7.82
feathered	nr	0.40	5.60	6.30	1.79	13.69
light standard	nr	1.00	14.00	13.50	4.13	31.63
standard	nr	1.20	16.80	15.55	4.85	37.20
selected standard	nr	1.40	19.60	25.30	6.74	51.64
heavy standard	nr	1.60	22.40	46.00	10.26	78.66
extra heavy standard	nr	2.00	28.00	57.50	12.83	98.33
petiolaris						
whip	nr	0.30	4.20	3.75	1.19	9.14
feathered	nr	0.40	5.60	8.05	2.05	15.70
light standard	nr	1.00	14.00	16.10	4.52	34.62
standard	nr	1.20	16.80	18.40	5.28	40.48
selected standard	nr	1.40	19.60	31.60	7.68	58.88
heavy standard	nr	1.60	22.40	51.75	11.12	85.27
extra heavy standard	nr	2.00	28.00	63.25	13.69	104.94

	Unit	Labour	Hours £	Mat'ls £	O & P £	Total £
Tree planting (cont'd)						
p. rubra						
whip	nr	0.30	4.20	3.75	1.19	9.14
feathered	nr	0.40	5.60	8.05	2.05	15.70
light standard	nr	1.00	14.00	16.10	4.52	34.62
standard	nr	1.20	16.80	18.40	5.28	40.48
selected standard	nr	1.40	19.60	31.60	7.68	58.88
heavy standard	nr	1.60	22.40	51.75	11.12	85.27
extra heavy standard	nr	2.00	28.00	63.25	13.69	104.94
CONTAINER GROWN TREES						
Excavate tree pit, fork bottom, plant tree, backfill with excavated material including organic manure (30% of soil by volume), water and surround with peat						
Acer (Maple)						
campestre carnival, size 6-8cm, pot 25 litre	nr	1.00	14.00	32.20	6.93	53.13
c.Postelense, size 6-8cm, pot 25 litre	nr	1.00	14.00	32.20	6.93	53.13
negundo elegans, size 6-8cm pot 25 litre	nr	1.00	14.00	29.90	6.59	50.49
p. Deborah, size 6-8cm, pot 25 litre	nr	1.00	14.00	27.60	6.24	47.84
p. Deborah, size 10-12cm, pot 45 litre	nr	1.40	19.60	64.40	12.60	96.60

	Unit	Labour	Hours £	Mat'ls £	O & P £	Total £
p. Drummondii, size 6-8cm, pot 25 litre	nr	1.00	14.00	27.60	6.24	47.84
p. Drummondii, size 10-12cm, pot 45 litre	nr	1.40	19.60	64.40	12.60	96.60
Aesculus (Chestnut)						
x carnea Briotii, size 6-8cm, pot 25 litre	nr	1.00	14.00	27.60	6.24	47.84
x carnea Briotii, size 10-12cm, pot 45 litre	nr	1.40	19.60	64.40	12.60	96.60
Alnus						
incarna Aurea, size 6-8cm, pot 25 litre	nr	1.00	14.00	29.90	6.59	50.49
incarna Aurea, size 10-12cm, pot 45 litre	nr	1.40	19.60	82.80	15.36	117.76
Betula (Birch)						
albosinensis China Ruby size 6-8cm, pot 25 litre	nr	1.00	14.00	32.20	6.93	53.13
nigra, size 10-12cm, pot 45 litre	nr	1.40	19.60	78.20	14.67	112.47
pendula, size 6-8cm, pot 25 litre	nr	1.00	14.00	21.85	5.38	41.23
pendula, size 10-12cm, pot 45 litre	nr	1.40	19.60	60.95	12.08	92.63
Laciniata, size 6-8cm, pot 25 litre	nr	1.00	14.00	32.20	6.93	53.13
Laciniata, size 10-12cm, pot 45 litre	nr	1.40	19.60	79.35	14.84	113.79

	Unit	Labour	Hours £	Mat'ls £	O & P £	Total £
Tree planting (cont'd)						
p. Youngii, size 6-8cm, pot 25 litre	nr	1.00	14.00	32.20	6.93	53.13
p. Youngii, size 10-12cm, pot 45 litre	nr	1.40	19.60	78.20	14.67	112.47
utilis var. jacquentmontii Doorenbos, size 6-8cm, pot 25 litre	nr	1.00	14.00	32.20	6.93	53.13
utilis var. jacquentmontii Doorenbos, size 10-12cm, pot 45 litre	nr	1.40	19.60	78.20	14.67	112.47
Fagus (Beech)						
sylvatica, size 6-8cm, pot 25 litre	nr	1.00	14.00	27.60	6.24	47.84
sylvatica, size 10-12cm, pot 45 litre	nr	1.40	19.60	102.35	18.29	140.24
s. Altropurperea, size 6-8cm, pot 25 litre	nr	1.00	14.00	32.20	6.93	53.13
s. Altropurperea, size 10-12cm, pot 45 litre	nr	1.40	19.60	102.35	18.29	140.24
Fraxinus (Ash)						
Augustifolia Raywood, size 6-8cm, pot 25 litre	nr	1.00	14.00	27.60	6.24	47.84
Augustifolia Raywood, size 10-12cm, pot 45 litre	nr	1.40	19.60	62.10	12.26	93.96
excelsior Jaspidea, size 6-8cm, pot 25 litre	nr	1.00	14.00	32.20	6.93	53.13
excelsior Jaspidea, size 10-12cm, pot 45 litre	nr	1.40	19.60	71.30	13.64	104.54

	Unit	Labour	Hours £	Mat'ls £	O & P £	Total £
Prunus (Cherry)						
Accolade, size 6-8cm, pot 25 litre	nr	1.00	14.00	27.60	6.24	47.84
avium Plena, size 6-8cm, pot 25 litre	nr	1.00	14.00	27.60	6.24	47.84
avium Plena, size 10-12cm, pot 45 litre	nr	1.40	19.60	62.10	12.26	93.96
Fragrant Cloud, size 6-8cm, pot 25 litre	nr	1.00	14.00	27.60	6.24	47.84
Kanzan, size 6-8cm, pot 25 litre	nr	1.00	14.00	27.60	6.24	47.84
Kanzan, size 10-12cm, pot 45 litre	nr	1.40	19.60	62.10	12.26	93.96
Pandora, size 6-8cm, pot 25 litre	nr	1.00	14.00	27.60	6.24	47.84
Royal Burgundy, size 6-8cm, pot 25 litre	nr	1.00	14.00	32.20	6.93	53.13
sargentii, size 6-8cm, pot 25 litre	nr	1.00	14.00	27.60	6.24	47.84
serrula, size 6-8cm, pot 25 litre	nr	1.00	14.00	27.60	6.24	47.84
x s. Autumnalis Rosea, size 6-8cm, pot 25 litre	nr	1.00	14.00	27.60	6.24	47.84
x s. Pendula Rubra, size 6-8cm, pot 25 litre	nr	1.00	14.00	27.60	6.24	47.84
Umineko, size 6-8cm, pot 25 litre	nr	1.00	14.00	27.60	6.24	47.84
Umineko, size 10-12cm, pot 45 litre	nr	1.40	19.60	66.70	12.95	99.25
x yeodensis, size 6-8cm, pot 25 litre	nr	1.00	14.00	27.60	6.24	47.84
x y Shidare-yoshino, size 6-8cm, pot 25 litre	nr	1.00	14.00	27.60	6.24	47.84

	Unit	Labour	Hours £	Mat'ls £	O & P £	Total £
Tree planting (cont'd)						
Quercus (Oak)						
ilex, size 6-8cm, pot 25 litre	nr	1.00	14.00	32.20	6.93	53.13
ilex, size 10-12cm, pot 45 litre	nr	1.40	19.60	72.45	13.81	105.86
robur f. fastigiata, size 10-12cm, pot 45 litre	nr	1.40	19.60	106.95	18.98	145.53
Sorbus (Whitebeam)						
aria Lutescens, size 6-8cm, pot 25 litre	nr	1.00	14.00	27.60	6.24	47.84
aria Lutescens, size 10-12cm, pot 45 litre	nr	1.40	19.60	62.10	12.26	93.96
majestica, size 6-8cm, pot 25 litre	nr	1.00	14.00	27.60	6.24	47.84
majestica, size 10-12cm, pot 45 litre	nr	1.40	19.60	62.10	12.26	93.96
a. var. xanthocarpa, size 6-8cm, pot 25 litre	nr	1.00	14.00	27.60	6.24	47.84
commixta Embley, size 6-8cm, pot 25 litre	nr	1.00	14.00	27.60	6.24	47.84
hupehensis, size 6-8cm, pot 25 litre	nr	1.00	14.00	27.60	6.24	47.84
hupehensis, size 10-12cm, pot 45 litre	nr	1.40	19.60	71.30	13.64	104.54
Leonard Messel, size 6-8cm, pot 25 litre	nr	1.00	14.00	27.60	6.24	47.84
sargentiana, size 6-8cm, pot 25 litre	nr	1.00	14.00	27.60	6.24	47.84
Sunshine, size 6-8cm, pot 25 litre	nr	1.00	14.00	27.60	6.24	47.84

CONIFERS

Excavate tree pit, fork bottom, plant conifer, backfill with excavated material including organic manure (30% of soil by volume), water and surround with peat

	Unit	Labour	Hours £	Mat'ls £	O & P £	Total £
Abies (Fir)						
concolor, height 60cm, pot 3 litre	nr	0.30	4.20	4.35	1.28	9.83
Cedrus (Cedar)						
antlantica, height 175cm, pot 35 litre	nr	1.00	14.00	63.25	11.59	88.84
antlantica, height 200cm, pot 50 litre	nr	1.00	14.00	92.00	15.90	121.90
antlantica, height 250cm, pot 75 litre	nr	1.00	14.00	138.00	22.80	174.80
deodara, height 175cm, pot 30 litre	nr	1.00	14.00	51.75	9.86	75.61
deodara, height 200cm, pot 35 litre	nr	1.00	14.00	74.75	13.31	102.06
deodara, height 250cm, pot 50 litre	nr	1.00	14.00	92.00	15.90	121.90
libani, height 175cm, pot 30 litre	nr	1.00	14.00	51.75	9.86	75.61
libani, height 200cm, pot 35 litre	nr	1.00	14.00	74.75	13.31	102.06
libani, height 250cm, pot 50 litre	nr	1.00	14.00	92.00	15.90	121.90

	Unit	Labour	Hours £	Mat'ls £	O & P £	Total £
Tree planting (cont'd)						
Chamaecyparis (Cypress)						
lawsoniana, height 80cm, pot 3 litre	nr	0.30	4.20	4.35	1.28	9.83
I. Ellwoodii, height 40cm, pot 2 litre	nr	0.30	4.20	4.35	1.28	9.83
I. Fletcherii, height 80cm, pot 3 litre	nr	0.30	4.20	4.35	1.28	9.83
I. stardust, height 80cm, pot 3 litre	nr	0.30	4.20	4.35	1.28	9.83
Crytopmeria (Cedar)						
japonica, height 60cm, pot 3 litre	nr	0.30	4.20	5.40	1.44	11.04
leylandii, height 80cm, pot 3 litre	nr	0.30	4.20	3.45	1.15	8.80
leylandii, height 100cm, pot 5 litre	nr	0.30	4.20	6.60	1.62	12.42
leylandii, height 125cm, pot 10 litre	nr	0.30	4.20	9.20	2.01	15.41
leylandii, height 150cm, pot 10 litre	nr	0.30	4.20	11.50	2.36	18.06
leylandii, height 175cm, pot 13 litre	nr	0.30	4.20	17.25	3.22	24.67
Cupressus (Cypress)						
macrocarpa Goldcrest, height 125cm, pot 10 litre	nr	0.30	4.20	27.60	4.77	36.57
macrocarpa Goldcrest, height 150cm, pot 15 litre	nr	0.30	4.20	46.00	7.53	57.73

	Unit	Labour	Hours £	Mat'ls £	O & P £	Total £
macrocarpa Goldcrest, height 200cm, pot 25 litre	nr	0.30	4.20	69.00	10.98	84.18
macrocarpa Goldcrest, height 250cm, pot 30 litre	nr	0.30	4.20	92.00	14.43	110.63
Larix (Larch)						
decidua, height 80cm, pot 3 litre	nr	0.30	4.20	4.40	1.29	9.89
x eurolepis, height 80cm, pot 3 litre	nr	0.30	4.20	4.40	1.29	9.89
kaempferi, height 80cm, pot 3 litre	nr	0.30	4.20	4.40	1.29	9.89
Pinus (Pine)						
nigra subsp laricio, height 80cm, pot 3 litre	nr	0.30	4.20	4.00	1.23	9.43
nigra subsp laricio, height 100cm, pot 10 litre	nr	0.30	4.20	11.50	2.36	18.06
nigra subsp laricio, height 125cm, pot 10 litre	nr	0.30	4.20	12.65	2.53	19.38
mugo, height 30cm, pot 2 litre	nr	0.30	4.20	5.20	1.41	10.81
mugo, height 50cm, pot 10 litre	nr	0.30	4.20	17.25	3.22	24.67

	Unit	Labour	Hours £	Mat'ls £	O & P £	Total £
Taxus (Yew)						
baccata, height 60cm, pot 3 litre	nr	0.30	4.20	5.20	1.41	10.81
baccata, height 80cm, pot 5 litre	nr	0.30	4.20	8.05	1.84	14.09
baccata, height 100cm, pot 10 litre	nr	0.30	4.20	17.25	3.22	24.67
baccata, height 125cm, pot 20 litre	nr	0.30	4.20	34.50	5.81	44.51

SHRUBS

The following descriptions refer to dimensions and sizes of shrubs as as set out below.

Dwarf: 20 to 60cm high
Small: 60 to 120cm high
Medium: 120 to 200cm high
Large: over 200cm high

Form planting hole in cultivated area, place shrub in hole, backfill, water and surround with peat

	Unit	Labour	Hours £	Mat'ls £	O & P £	Total £
Acer (Maple)						
campestre, dwarf, pot 3 litre	nr	0.20	2.80	2.30	0.77	5.87
campestre, medium, pot 10 litre	nr	0.40	5.60	11.50	2.57	19.67
p. f. atropurpureum, dwarf, pot 3 litre	nr	0.20	2.80	12.65	2.32	17.77

	Unit	Labour	Hours £	Mat'ls £	O & P £	Total £
p. f. atropurpureum, small, pot 10 litre	nr	0.30	4.20	28.75	4.94	37.89
p. Bloodgood, small, pot 3 litre	nr	0.30	4.20	12.65	2.53	19.38
Sango-kaku, dwarf, pot 3 litre	nr	0.20	2.80	12.65	2.32	17.77
Sango-kaku, small, pot 3 litre	nr	0.30	4.20	28.75	4.94	37.89
shirawanum, dwarf pot 3 litre	nr	0.20	2.80	12.65	2.32	17.77
Buddleja (Buddleia)						
davidii Black Knight, dwarf, pot 3 litre	nr	0.20	2.80	4.00	1.02	7.82
davidii Black Knight, small, pot 10 litre	nr	0.30	4.20	12.65	2.53	19.38
d. Harlequin, dwarf, pot 3 litre	nr	0.20	2.80	4.00	1.02	7.82
d. Harlequin, small, pot 10 litre	nr	0.30	4.20	12.65	2.53	19.38
d. Royal Red, dwarf, pot 3 litre	nr	0.20	2.80	4.00	1.02	7.82
d. Royal Red, small, pot 10 litre	nr	0.30	4.20	12.65	2.53	19.38
d. White Profusion, dwarf, pot 3 litre	nr	0.20	2.80	4.00	1.02	7.82
d. White Profusion, small, pot 10 litre	nr	0.30	4.20	12.65	2.53	19.38
Pink Delight, dwarf, pot 3 litre	nr	0.20	2.80	4.00	1.02	7.82
Pink Delight, small, pot 10 litre	nr	0.30	4.20	12.65	2.53	19.38

	Unit	Labour	Hours £	Mat'ls £	O & P £	Total £
Camellia (Camellia)						
japonica Adolphe Audusson, dwarf, pot 3 litre	nr	0.20	2.80	5.75	1.28	9.83
japonica Adolphe Audusson, small, pot 3 litre	nr	0.30	4.20	7.50	1.76	13.46
japonica Adolphe Audusson, medium, pot 10 litre	nr	0.40	5.60	16.10	3.26	24.96
j. Lady Vansittart, dwarf, pot 3 litre	nr	0.20	2.80	5.75	1.28	9.83
j. Lady Vansittart, small, pot 3 litre	nr	0.30	4.20	7.50	1.76	13.46
j. Lady Vansittart, medium, pot 10 litre	nr	0.40	5.60	16.10	3.26	24.96
Leonard Messel, dwarf, pot 3 litre	nr	0.20	2.80	5.75	1.28	9.83
Leonard Messel, small, pot 3 litre	nr	0.30	4.20	7.50	1.76	13.46
Leonard Messel, medium, pot 10 litre	nr	0.40	5.60	16.10	3.26	24.96
x w. Debbie, dwarf, pot 3 litre	nr	0.20	2.80	5.75	1.28	9.83
Leonard Messel, small, pot 3 litre	nr	0.30	4.20	7.50	1.76	13.46
Leonard Messel, medium, pot 10 litre	nr	0.40	5.60	16.10	3.26	24.96
Cytisus (Broom)						
battandieri, dwarf, pot 3 litre	nr	0.20	2.80	7.50	1.55	11.85
battandieri, small, pot 10 litre	nr	0.30	4.20	16.10	3.05	23.35

	Unit	Labour	Hours £	Mat'ls £	O & P £	Total £
burkwoodii, dwarf, pot 3 litre	nr	0.20	2.80	4.00	1.02	7.82
burkwoodii, small, pot 10 litre	nr	0.30	4.20	12.65	2.53	19.38
Fulgens, dwarf, pot 3 litre	nr	0.20	2.80	4.00	1.02	7.82
Fulgens, small, pot 10 litre	nr	0.30	4.20	12.65	2.53	19.38
Goldfinch, dwarf, pot 3 litre	nr	0.20	2.80	4.00	1.02	7.82
Goldfinch, small, pot 10 litre	nr	0.30	4.20	12.65	2.53	19.38
Lena, dwarf, pot 3 litre	nr	0.20	2.80	4.00	1.02	7.82
Lena, small, pot 10 litre	nr	0.30	4.20	12.65	2.53	19.38
x p. Allgold, dwarf, pot 3 litre	nr	0.20	2.80	4.00	1.02	7.82
x p. Allgold, small, pot 10 litre	nr	0.30	4.20	12.65	2.53	19.38
Zeelandia, dwarf, pot 3 litre	nr	0.20	2.80	4.00	1.02	7.82
Fagus (Beech)						
sylvatica, dwarf, pot 3 litre	nr	0.20	2.80	2.30	0.77	5.87
sylvatica, dwarf, pot 10 litre	nr	0.30	4.20	11.50	2.36	18.06
Hydrangea						
arborescens Annabelle, dwarf, pot 3 litre	nr	0.20	2.80	4.00	1.02	7.82
arborescens Annabelle, small, pot 10 litre	nr	0.30	4.20	12.65	2.53	19.38

	Unit	Labour	Hours £	Mat'ls £	O & P £	Total £
Shrub planting (cont'd)						
a Villosa Group, dwarf, pot 3 litre	nr	0.20	2.80	5.76	1.28	9.84
a Villosa Group, small, pot 10 litre	nr	0.30	4.20	12.65	2.53	19.38
m Mariesii Grandiflora, dwarf, pot 3 litre	nr	0.20	2.80	4.00	1.02	7.82
m Mariesii Grandiflora, small, pot 10 litre	nr	0.30	4.20	12.65	2.53	19.38
m Mowe, dwarf, pot 3 litre	nr	0.20	2.80	4.00	1.02	7.82
m Mowe, dwarf, pot 10 litre	nr	0.30	4.20	12.65	2.53	19.38
Preziosa, dwarf, pot 3 litre	nr	0.20	2.80	4.00	1.02	7.82
Preziosa, dwarf, pot 10 litre	nr	0.30	4.20	12.65	2.53	19.38
q. Snow Queen, dwarf, pot 3 litre	nr	0.20	2.80	5.75	1.28	9.83
q. Snow Queen, small, pot 10 litre	nr	0.30	4.20	12.65	2.53	19.38
s. Miranda, dwarf, pot 3 litre	nr	0.20	2.80	5.75	1.28	9.83
Lavandula (Lavender)						
augustifolia, dwarf, pot 2 litre	nr	0.20	2.80	2.30	0.77	5.87
a. Munstead, dwarf, pot 2 litre	nr	0.20	2.80	2.30	0.77	5.87
a. Twickel Purple, dwarf, pot 2 litre	nr	0.20	2.80	2.30	0.77	5.87
stoechas, dwarf, pot 2 litre	nr	0.20	2.80	2.30	0.77	5.87

	Unit	Labour	Hours £	Mat'ls £	O & P £	Total £
Ligustrum (Privet)						
japonicum, dwarf, pot 3 litre	nr	0.20	2.80	4.60	1.11	8.51
japonicum, small, pot 10 litre	nr	0.30	4.20	12.65	2.53	19.38
ovalifolium, dwarf, pot 3 litre	nr	0.20	2.80	2.30	0.77	5.87
ovalifolium, small, pot 10 litre	nr	0.30	4.20	11.50	2.36	18.06
o. Aureum, dwarf, pot 3 litre	nr	0.20	2.80	3.45	0.94	7.19
o. Aureum, small, pot 10 litre	nr	0.30	4.20	12.65	2.53	19.38
vulgare, dwarf, pot 3 litre	nr	0.20	2.80	2.30	0.77	5.87
vulgare, small, pot 10 litre	nr	0.30	4.20	11.50	2.36	18.06
Lonicera (Honeysuckle)						
fragrantissima, dwarf, pot 3 litre	nr	0.20	2.80	5.75	1.28	9.83
fragrantissima, small, pot 10 litre	nr	0.30	4.20	12.65	2.53	19.38
nitida, dwarf, pot 3 litre	nr	0.20	2.80	2.30	0.77	5.87
nitida, small, pot 10 litre	nr	0.30	4.20	12.65	2.53	19.38
n. Baggesen's Gold, dwarf, pot 3 litre	nr	0.20	2.80	3.45	0.94	7.19
n. Baggesen's Gold, small, pot 10 litre	nr	0.30	4.20	12.65	2.53	19.38
pileata, dwarf, pot 3 litre	nr	0.20	2.80	2.30	0.77	5.87

	Unit	Labour	Hours £	Mat'ls £	O & P £	Total £
Shrub planting (cont'd)						
pileata, small, pot 10 litre	nr	0.20	2.80	2.30	0.77	5.87
syringantha, dwarf, pot 3 litre	nr	0.20	2.80	5.75	1.28	9.83
Magnolia						
x brooklynensis, dwarf, pot 3 litre	nr	0.20	2.80	10.35	1.97	15.12
g. Galisonniere, dwarf, pot 3 litre	nr	0.20	2.80	10.35	1.97	15.12
g. Galisonniere, small, pot 10 litre	nr	0.30	4.20	25.85	4.51	34.56
Heaven Scent, dwarf, pot 3 litre	nr	0.20	2.80	10.35	1.97	15.12
Heaven Scent, small, pot 10 litre	nr	0.30	4.20	25.85	4.51	34.56
liliflora Nigra, dwarf, pot 3 litre	nr	0.20	2.80	10.35	1.97	15.12
liliflora Nigra, small, pot 10 litre	nr	0.30	4.20	25.85	4.51	34.56
x soulangeana, dwarf, pot 3 litre	nr	0.20	2.80	10.35	1.97	15.12
x soulangeana, small, pot 10 litre	nr	0.30	4.20	25.85	4.51	34.56
stellata, dwarf, pot 3 litre	nr	0.20	2.80	10.35	1.97	15.12
stellata, small, pot 10 litre	nr	0.30	4.20	25.85	4.51	34.56
s. Royal Star, dwarf, pot 3 litre	nr	0.20	2.80	10.35	1.97	15.12
Sundance, dwarf, pot 3 litre	nr	0.20	2.80	10.35	1.97	15.12

	Unit	Labour	Hours £	Mat'ls £	O & P £	Total £
Prunus (Cherry)						
cerasifera Nigra, dwarf, pot 3 litre	nr	0.20	2.80	5.75	1.28	9.83
cerasifera Nigra, small, pot 10 litre	nr	0.30	4.20	12.65	2.53	19.38
laurocerasus, dwarf, pot 3 litre	nr	0.20	2.80	2.30	0.77	5.87
laurocerasus, small, pot 10 litre	nr	0.30	4.20	12.65	2.53	19.38
l. Cherry Brandy, dwarf, pot 3 litre	nr	0.20	2.80	4.60	1.11	8.51
l. Cherry Brandy, small, pot 10 litre	nr	0.30	4.20	12.65	2.53	19.38
l. Mount Vernon, dwarf, pot 3 litre	nr	0.20	2.80	4.60	1.11	8.51
l. Mount Vernon, small, pot 10 litre	nr	0.30	4.20	12.65	2.53	19.38
l. Schipkaensis, dwarf, pot 3 litre	nr	0.20	2.80	3.45	0.94	7.19
l. Schipkaensis, small, pot 10 litre	nr	0.30	4.20	12.65	2.53	19.38
l. Zabeliana, dwarf, pot 3 litre	nr	0.20	2.80	3.45	0.94	7.19
l. Zabeliana, small, pot 10 litre	nr	0.30	4.20	12.65	2.53	19.38
lusitanica, dwarf, pot 3 litre	nr	0.20	2.80	3.45	0.94	7.19
lusitanica, small, pot 10 litre	nr	0.30	4.20	12.65	2.53	19.38
l. Variegata, dwarf, pot 3 litre	nr	0.20	2.80	3.45	0.94	7.19
l. Variegata, small, pot 10 litre	nr	0.30	4.20	12.65	2.53	19.38

	Unit	Labour	Hours £	Mat'ls £	O & P £	Total £
Shrub planting (cont'd)						
spinosa, dwarf, pot 3 litre	nr	0.20	2.80	2.30	0.77	5.87
Pyracantha (Firethorn)						
Red column, dwarf, pot 2 litre	nr	0.20	2.80	2.90	0.86	6.56
Red column, small, pot 10 litre	nr	0.30	4.20	12.65	2.53	19.38
Mohave, dwarf, pot 2 litre	nr	0.20	2.80	2.90	0.86	6.56
Mohave, dwarf, pot 10 litre	nr	0.30	4.20	12.65	2.53	19.38
Orange Glow, dwarf, pot 2 litre	nr	0.20	2.80	3.30	0.92	7.02
Orange Glow, small, pot 10 litre	nr	0.30	4.20	12.65	2.53	19.38
Soleil d'Or, dwarf, pot 2 litre	nr	0.20	2.80	3.30	0.92	7.02
Soleil d'Or, small, pot 10 litre	nr	0.30	4.20	12.65	2.53	19.38
Sparkler, dwarf, pot 2 litre	nr	0.20	2.80	2.90	0.86	6.56
Teton, dwarf, pot 2 litre	nr	0.20	2.80	2.90	0.86	6.56
Salix (Willow)						
hastata Wehrhahnii, dwarf, pot 3 litre	nr	0.20	2.80	5.75	1.28	9.83
hastata Wehrhahnii, small, pot 10 litre	nr	0.30	4.20	12.65	2.53	19.38
helvetica, dwarf, pot 2 litre	nr	0.20	2.80	5.75	1.28	9.83

	Unit	Labour	Hours £	Mat'ls £	O & P £	Total £
helvetica, small, pot 10 litre	nr	0.30	4.20	12.65	2.53	19.38
viminalis, dwarf, pot 3 litre	nr	0.20	2.80	2.30	0.77	5.87
viminalis, small, pot 10 litre	nr	0.30	4.20	11.50	2.36	18.06
Syringa (Lilac)						
meyeri var.spontanea Palibin, dwarf, pot 3 litre	nr	0.20	2.80	6.60	1.41	10.81
meyeri var.spontanea Palibin, small, pot 10 litre	nr	0.30	4.20	16.10	3.05	23.35
pubescens. Subsp. microphylla Superba, dwarf, pot 3 litre	nr	0.20	2.80	6.60	1.41	10.81
pubescens. Subsp. microphylla Superba, dwarf, pot 10 litre	nr	0.30	4.20	16.10	3.05	23.35
v. Charles Joly, dwarf, pot 3 litre	nr	0.20	2.80	6.60	1.41	10.81
v. Charles Joly, small, pot 10 litre	nr	0.30	4.20	16.10	3.05	23.35
v. Madame Lemoine, dwarf, pot 3 litre	nr	0.20	2.80	6.60	1.41	10.81
v. Madame Lemoine, small, pot 10 litre	nr	0.30	4.20	16.10	3.05	23.35
v. Michael Buchner, small, pot 10 litre	nr	0.30	4.20	16.10	3.05	23.35

	Unit	Labour	Hours £	Mat'ls £	O & P £	Total £
Specimen shrubs						
The following descriptions refer to dimensions and sizes of shrubs as as set out below.						
	Dwarf: 20 to 60cm high Small: 60 to 120cm high Medium: 120 to 200cm high Large: over 200cm high					
Form planting hole in cultivated area, place shrub in hole, backfill, water and surround with peat						
Acer (Maple)						
p. Fireglow, medium, pot 25 litre	nr	0.60	8.40	69.00	11.61	89.01
p. Fireglow, medium, pot 50 litre	nr	1.00	14.00	115.00	19.35	148.35
p. Fireglow, large, pot 130 litre	nr	1.80	25.20	230.00	38.28	293.48
Betula (Birch)						
pendula, large, pot 55 litre	nr	1.00	14.00	48.90	9.44	72.34
Prunus (Cherry)						
laurocerasus, medium, pot 15 litre	nr	0.40	5.60	51.75	8.60	65.95
laurocerasus, medium, pot 18 litre	nr	0.40	5.60	69.00	11.19	85.79

	Unit	Labour	Hours £	Mat'ls £	O & P £	Total £
laurocerasus, medium, pot 25 litre	nr	0.60	8.40	103.50	16.79	128.69
laurocerasus, medium, pot 55 litre	nr	1.00	14.00	138.00	22.80	174.80
lusitanica, medium, pot 20 litre	nr	0.60	8.40	74.75	12.47	95.62
lusitanica, medium, pot 25 litre	nr	0.60	8.40	97.75	15.92	122.07

CLIMBERS

The following descriptions refer to dimensions and sizes of shrubs as as set out below.

Dwarf: 20 to 60cm high
Small: 60 to 120cm high
Medium: 120 to 200cm high
Large: over 200cm high

Form planting hole in cultivated area, place climber in hole, backfill, water and surround with peat

Clematis

	Unit	Labour	Hours £	Mat'ls £	O & P £	Total £
Blue Bird, small, pot 2 litre	nr	0.10	1.40	5.75	1.07	8.22
x durandi, small, pot 2 litre	nr	0.10	1.40	5.75	1.07	8.22
Huldine, small, pot 2 litre	nr	0.10	1.40	5.75	1.07	8.22
Jackmanii, small, pot 2 litre	nr	0.10	1.40	5.75	1.07	8.22

	Unit	Labour	Hours £	Mat'ls £	O & P £	Total £
Climber planting (cont'd)						
Laserstern, small, pot 2 litre	nr	0.10	1.40	5.75	1.07	8.22
Miss Bateman, small, pot 2 litre	nr	0.10	1.40	5.75	1.07	8.22
Nelly Moser, small, pot 2 litre	nr	0.10	1.40	5.75	1.07	8.22
Rouge Cardinal, small, pot 2 litre	nr	0.10	1.40	5.75	1.07	8.22
Sunset, small, pot 2 litre	nr	0.10	1.40	5.75	1.07	8.22
The President, small, pot 2 litre	nr	0.10	1.40	5.75	1.07	8.22
Vyvyan Pennell, small, pot 2 litre	nr	0.10	1.40	5.75	1.07	8.22
Wyevale, small, pot 2 litre	nr	0.10	1.40	5.75	1.07	8.22
Clematis Species						
Alba Luxurians, small, pot 2 litre	nr	0.10	1.40	5.75	1.07	8.22
armandii, small, pot 2 litre	nr	0.10	1.40	10.35	1.76	13.51
a. Apple Blossom, small, pot 2 litre	nr	0.10	1.40	5.75	1.07	8.22
flamula, small, pot 2 litre	nr	0.10	1.40	5.75	1.07	8.22
montana, small, pot 2 litre	nr	0.10	1.40	5.75	1.07	8.22
m. Grandiflora, small, pot 2 litre	nr	0.10	1.40	5.75	1.07	8.22
m. var. rubens, small, pot 2 litre	nr	0.10	1.40	5.75	1.07	8.22

	Unit	Labour	Hours £	Mat'ls £	O & P £	Total £
montana var. sericea, small, pot 2 litre	nr	0.10	1.40	5.75	1.07	8.22
o. Orange Peel, small, pot 2 litre	nr	0.10	1.40	5.75	1.07	8.22
tangutica, small, pot 2 litre	nr	0.10	1.40	5.75	1.07	8.22
Venosa Violacea, small pot 2 litre	nr	0.10	1.40	5.75	1.07	8.22
vitalba, small, pot 2 litre	nr	0.10	1.40	5.75	1.07	8.22
viticella, small, pot 2 litre	nr	0.10	1.40	5.75	1.07	8.22

Hedera (Ivy)

	Unit	Labour	Hours £	Mat'ls £	O & P £	Total £
canariensis, dwarf, pot 2 0.5 litre	nr	0.10	1.40	1.40	0.42	3.22
canariensis, small, pot 2 litre	nr	0.10	1.40	4.30	0.86	6.56
c. ravensholt, dwarf, pot 0.5 litre	nr	0.10	1.40	1.40	0.42	3.22
c. ravensholt, small, pot 2 litre	nr	0.10	1.40	4.30	0.86	6.56
c. dentata, dwarf, pot 0.5 litre	nr	0.10	1.40	1.40	0.42	3.22
c. dentata, small, pot 2 litre	nr	0.10	1.40	4.30	0.86	6.56
c. Sulphur Heart, dwarf, pot 0.5 litre	nr	0.10	1.40	1.40	0.42	3.22
c. Sulphur Heart, small, pot 2 litre	nr	0.10	1.40	4.30	0.86	6.56
helix, dwarf, pot 0.5 litre	nr	0.10	1.40	1.40	0.42	3.22
helix, small, pot 2 litre	nr	0.10	1.40	4.30	0.86	6.56

	Unit	Labour	Hours £	Mat'ls £	O & P £	Total £
Climber planting (cont'd)						
h. buttercup, dwarf, pot 0.5 litre	nr	0.10	1.40	1.40	0.42	3.22
h. buttercup, small, pot 2 litre	nr	0.10	1.40	4.30	0.86	6.56
h. Goldchild, dwarf, pot 0.5 litre	nr	0.10	1.40	1.40	0.42	3.22
h. Goldchild, small, pot 2 litre	nr	0.10	1.40	4.30	0.86	6.56
c. Green Ripple, dwarf, pot 0.5 litre	nr	0.10	1.40	1.40	0.42	3.22
c. Green Ripple, small, pot 2 litre	nr	0.10	1.40	4.30	0.86	6.56
h. Ivalace, dwarf, pot 0.5 litre	nr	0.10	1.40	1.40	0.42	3.22
h. Ivalace, small, pot 2 litre	nr	0.10	1.40	4.30	0.86	6.56
h. Kolibri, dwarf, pot 0.5 litre	nr	0.10	1.40	1.40	0.42	3.22
h. Kolibri, small, pot 2 litre	nr	0.10	1.40	4.30	0.86	6.56
c. Light Fingers, dwarf, pot 0.5 litre	nr	0.10	1.40	1.40	0.42	3.22
c. Light Fingers, small, pot 2 litre	nr	0.10	1.40	4.30	0.86	6.56
c. Little Diamond, dwarf, pot 0.5 litre	nr	0.10	1.40	1.40	0.42	3.22
c. Little Diamond, small, pot 2 litre	nr	0.10	1.40	4.20	0.84	6.44
h. Luzii, dwarf, pot 0.5 litre	nr	0.10	1.40	1.40	0.42	3.22
h. Luzii, small, pot 2 litre	nr	0.10	1.40	1.40	0.42	3.22

	Unit	Labour	Hours £	Mat'ls £	O & P £	Total £
Hydrangea						
anomala ssp. petiolaris, dwarf, pot 3 litre	nr	0.10	1.40	5.75	1.07	8.22
anomala ssp. petiolaris, small, pot 10 litre	nr	0.20	2.80	12.65	2.32	17.77
seemannii, small, pot pot 2 litre	nr	0.10	1.40	1.40	0.42	3.22
Jasminium (Jasmine)						
humile Revolutum, dwarf, pot 2 litre	nr	0.10	1.40	4.60	0.90	6.90
nudiflorum, dwarf, pot pot 2 litre	nr	0.20	2.80	4.60	1.11	8.51
nudiflorum, small, pot pot 10 litre	nr	0.20	2.80	12.65	2.32	17.77
o. aureum, small, pot pot 2 litre	nr	0.10	1.40	4.60	0.90	6.90
o. Fiona Sunrise, small, pot 2 litre	nr	0.10	1.40	5.75	1.07	8.22
o. Inverleith, small, pot 2 litre	nr	0.10	1.40	4.60	0.90	6.90
x. stephanense, small, pot 2 litre	nr	0.10	1.40	4.60	0.90	6.90
Lonicera (Honeysuckle)						
x. brownii, small, pot 2 litre	nr	0.10	1.40	4.60	0.90	6.90
x. heckrotii, small, pot 2 litre	nr	0.10	1.40	4.60	0.90	6.90
henryi, small, pot 2 litre	nr	0.10	1.40	4.60	0.90	6.90
x. italica, small, pot 2 litre	nr	0.10	1.40	4.60	0.90	6.90

	Unit	Labour	Hours £	Mat'ls £	O & P £	Total £
Climber planting (cont'd)						
japonica Aureoreticulata, small, pot 2 litre	nr	0.10	1.40	4.60	0.90	6.90
j. Halliana, small, pot 2 litre	nr	0.10	1.40	4.60	0.90	6.90
periclymenum, small, pot 2 litre	nr	0.10	1.40	4.60	0.90	6.90
p. Belgica, small, pot 2 litre	nr	0.10	1.40	4.60	0.90	6.90
sempervirens, small, pot 2 litre	nr	0.10	1.40	4.60	0.90	6.90
Passiflora (Passion flower)						
caerulea, small, pot 2 litre	nr	0.10	1.40	4.60	0.90	6.90
c. rubra, small, pot 2 litre	nr	0.10	1.40	4.60	0.90	6.90
Rosa (Rose)						
banksiae Lutea, small, pot 4 litre	nr	0.15	2.10	4.90	1.05	8.05
Mermaid, small, pot 4 litre	nr	0.15	2.10	4.90	1.05	8.05
Wisteria						
Burford, small, pot 3 litre	nr	0.15	2.10	9.20	1.70	13.00
Caroline, small, pot 3 litre	nr	0.15	2.10	9.20	1.70	13.00
Caroline, medium, pot 10 litre	nr	0.20	2.80	17.25	3.01	23.06

	Unit	Labour	Hours £	Mat'ls £	O & P £	Total £
floribunda Alba, small, pot 3 litre	nr	0.15	2.10	9.20	1.70	13.00
floribunda Alba, medium, pot 10 litre	nr	0.20	2.80	17.25	3.01	23.06
f. Domimo, small, pot 3 litre	nr	0.15	2.10	9.20	1.70	13.00
f. Domimo, medium, pot 10 litre	nr	0.20	2.80	17.25	3.01	23.06
f. Kuchi-beni, small, pot 3 litre	nr	0.15	2.10	9.20	1.70	13.00
f. Lawrence, small, pot 3 litre	nr	0.15	2.10	9.20	1.70	13.00
f. Multijuga, small, pot 3 litre	nr	0.15	2.10	9.20	1.70	13.00
f. Rosea, small, pot 3 litre	nr	0.15	2.10	9.20	1.70	13.00
f. Rosea, medium, pot 10 litre	nr	0.20	2.80	17.25	3.01	23.06
sinensis, small, pot 3 litre	nr	0.15	2.10	9.20	1.70	13.00
s. Alba, small, pot 3 litre	nr	0.15	2.10	9.20	1.70	13.00
s. Alba, medium, pot 10 litre	nr	0.20	2.80	17.25	3.01	23.06
s. Amethyst, small, pot 3 litre	nr	0.15	2.10	9.20	1.70	13.00
s. Amethyst, medium, pot 10 litre	nr	0.20	2.80	17.25	3.01	23.06
s. Blue Sapphire, small, pot 3 litre	nr	0.15	2.10	9.20	1.70	13.00
s. Prolific, small, pot 3 litre	nr	0.15	2.10	9.20	1.70	13.00
s. Prolific, medium, pot 10 litre	nr	0.20	2.80	17.25	3.01	23.06

	Unit	Labour	Hours £	Mat'ls £	O & P £	Total £
HERBACEOUS PLANTS						
Form planting hole in cultivated area, place plant in hole, backfill, water and surround with peat						
Anenome						
hupehensis, 750mm high, pot 2 litre	nr	0.15	2.10	3.45	0.83	6.38
h. var. japonica, 1000mm high, pot 2 litre	nr	0.15	2.10	3.45	0.83	6.38
h. Splendens, 800mm high, pot 2 litre	nr	0.15	2.10	3.45	0.83	6.38
h. Elegans, 800mm high, pot 2 litre	nr	0.15	2.10	3.45	0.83	6.38
x h. Robustissima, 1200mm high, pot 2 litre	nr	0.15	2.10	3.45	0.83	6.38
x h. September, 750mm high, pot 2 litre	nr	0.15	2.10	3.45	0.83	6.38
x h. Whirlwind, 900mm high, pot 2 litre	nr	0.15	2.10	3.45	0.83	6.38
sylvestis Macrantha, 250mm high, pot 2 litre	nr	0.15	2.10	3.45	0.83	6.38
Campanula (Bell flower)						
carpatica, 200mm high, pot 1 litre	nr	0.10	1.40	2.30	0.56	4.26
c. White Clips, 200mm high, pot 1 litre	nr	0.10	1.40	2.30	0.56	4.26

	Unit	Labour	Hours £	Mat'ls £	O & P £	Total £
cochlearifokia, 100mm high, pot 1 litre	nr	0.10	1.40	1.15	0.38	2.93
c. White Baby, 100mm high, pot 0.5 litre	nr	0.10	1.40	1.15	0.38	2.93
g. var.alba, 300mm high, pot 0.5 litre	nr	0.10	1.40	1.75	0.47	3.62
Kent Belle, 600mm high, pot 2 litre	nr	0.15	2.10	3.15	0.79	6.04
lactiflora Loddon Anna, 1200 high, pot 0.5 litre	nr	0.10	1.40	1.15	0.38	2.93
lactiflora Loddon Anna, 1500 high, pot 2 litre	nr	0.15	2.10	2.30	0.66	5.06
I. Violet, 1300mm high pot 0.5 litre	nr	0.10	1.40	1.15	0.38	2.93
I. Violet, 1300mm high pot 2 litre	nr	0.15	2.10	3.15	0.79	6.04
I. White Pouffe, 250mm high, pot 2 litre	nr	0.15	2.10	2.30	0.66	5.06
persificolia Alba, 900 high, pot 0.5 litre	nr	0.15	2.10	1.15	0.49	3.74
p. Caerulea Plena, 750mm high, pot 0.5 litre	nr	0.15	2.10	1.15	0.49	3.74
p. Telham Beauty, 900mm high, pot 2 litre	nr	0.15	2.10	1.15	0.49	3.74
p. Stella, 250mm high, pot 1 litre	nr	0.15	2.10	2.30	0.66	5.06
punctata, 300mm high, pot 2 litre	nr	0.15	2.10	2.30	0.66	5.06
takesimana, 600mm high, pot 2 litre	nr	0.15	2.10	2.30	0.66	5.06
I. Elizabeth, 600mm high, pot 2 litre	nr	0.15	2.10	3.15	0.79	6.04

	Unit	Labour	Hours £	Mat'ls £	O & P £	Total £
Herbaceous plants planting (cont'd)						
Delphinium						
Astolat Group, 1500mm high, pot 0.5 litre	nr	0.10	1.40	1.15	0.38	2.93
Astolat Group, 1500mm high, pot 2 litre	nr	0.15	2.10	2.90	0.75	5.75
Black Knight Group, 1800mm high, pot 0.5 litre	nr	0.10	1.40	1.15	0.38	2.93
Black Knight Group, 1800mm high, pot 2 litre	nr	0.15	2.10	2.90	0.75	5.75
Blue Bird Group, 1500mm high, pot 0.5 litre	nr	0.10	1.40	1.15	0.38	2.93
Blue Bird Group, 1500mm high, pot 2 litre	nr	0.15	2.10	2.90	0.75	5.75
Galahad Group, 1500mm high, pot 0.5 litre	nr	0.10	1.40	1.15	0.38	2.93
Galahad Group, 1500mm high, pot 2 litre	nr	0.15	2.10	2.90	0.75	5.75
grandiflorum, 400mm high, pot 0.5 litre	nr	0.10	1.40	1.15	0.38	2.93
grandiflorum, 400mm high, pot 2 litre	nr	0.15	2.10	2.90	0.75	5.75
King Arthur Group, 1500mm high, pot 0.5 litre	nr	0.10	1.40	1.15	0.38	2.93
King Arthur Group, 1500mm high, pot 2 litre	nr	0.15	2.10	2.90	0.75	5.75

	Unit	Labour	Hours £	Mat'ls £	O & P £	Total £
Summer Skies Group, 1500mm high, pot 0.5 litre	nr	0.10	1.40	1.15	0.38	2.93
Summer Skies Group, 1500mm high, pot 2 litre	nr	0.15	2.10	2.90	0.75	5.75
Geranium						
Anne Thomson, 600mm high, pot 2 litre	nr	0.15	2.10	3.15	0.79	6.04
Bertie Crug, 200mm high, pot 2 litre	nr	0.15	2.10	3.15	0.79	6.04
x cantabrigiense, 250mm high, pot 2 litre	nr	0.15	2.10	2.15	0.64	4.89
cinerum Ballerina, 100mm high, pot 0.5 litre	nr	0.10	1.40	1.80	0.48	3.68
cinerum Ballerina, 100mm high, pot 2 litre	nr	0.10	1.40	3.75	0.77	5.92
c. var. subcaulescens, Splendens, 150mm high, pot 0.5 litre	nr	0.10	1.40	1.80	0.48	3.68
c. var. subcaulescens, Splendens, 150mm high, pot 2 litre	nr	0.15	2.10	3.75	0.88	6.73
c. Kashmir Purple, 300mm high, pot 2 litre	nr	0.15	2.10	2.90	0.75	5.75
c. Kashmir White, 300mm high, pot 2 litre	nr	0.15	2.10	2.90	0.75	5.75
Johnson's Blue, 150mm high, pot 0.5 litre	nr	0.10	1.40	1.50	0.44	3.34
Johnson's Blue, 150mm high, pot 2 litre	nr	0.15	2.10	2.90	0.75	5.75

	Unit	Labour	Hours £	Mat'ls £	O & P £	Total £
Herbaceous plants planting (cont'd)						
macrorrhizum Album, 300mm high, pot 0.5 litre	nr	0.10	1.40	1.70	0.47	3.57
macrorrhizum Album, 300mm high, pot 2 litre	nr	0.15	2.10	3.15	0.79	6.04
m. Variegatum, 300mm high, pot 2 litre	nr	0.15	2.10	3.15	0.79	6.04
m. magnificum, 600mm high, pot 2 litre	nr	0.15	2.10	2.15	0.64	4.89
x. oxonianum, 450mm high, pot 2 litre	nr	0.15	2.10	2.15	0.64	4.89
x o. Wargrove Pink, 400mm high, pot 0.5 litre	nr	0.10	1.40	1.15	0.38	2.93
x o. Wargrove Pink, 400mm high, pot 2 litre	nr	0.15	2.10	2.15	0.64	4.89
Patricia, 750mm high, pot 2 litre	nr	0.15	2.10	3.15	0.79	6.04
phaeum, 800mm high, pot 2 litre	nr	0.15	2.10	3.15	0.79	6.04
p. Samobor, 400mm high, pot 2 litre	nr	0.15	2.10	3.15	0.79	6.04
pstilostemon, 1200mm high, pot 2 litre	nr	0.15	2.10	3.15	0.79	6.04
renardi, 250mm high, pot 2 litre	nr	0.15	2.10	3.15	0.79	6.04
x r. Russell Prichard, 200mm high, pot 0.5 litre	nr	0.10	1.40	1.80	0.48	3.68

	Unit	Labour	Hours £	Mat'ls £	O & P £	Total £
x r. Russell Prichard, 200mm high, pot 2 litre	nr	0.15	2.10	3.75	0.88	6.73
Salome, 450mm high, pot 2 litre	nr	0.15	2.10	3.15	0.79	6.04
s. Album, 300mm high, pot 2 litre	nr	0.15	2.10	3.15	0.79	6.04
s. Max Frei, 200mm high, pot 2 litre	nr	0.15	2.10	3.15	0.79	6.04
Spinners, 900mm high, pot 2 litre	nr	0.15	2.10	3.15	0.79	6.04
Sue Crug, 300mm high, pot 2 litre	nr	0.15	2.10	3.15	0.79	6.04
s. Mayflower, 300mm high, pot 2 litre	nr	0.15	2.10	3.15	0.79	6.04
Westray, 300mm high, pot 2 litre	nr	0.15	2.10	3.15	0.79	6.04
Gypsophila						
paniculata Bristol Fairy, 1200mm high, pot 0.5 litre	nr	0.10	1.40	1.45	0.43	3.28
paniculata Bristol Fairy, 1200mm high, pot 2 litre	nr	0.15	2.10	3.45	0.83	6.38
p.Flamingo, 900mm high, pot 0.5 litre	nr	0.10	1.40	1.45	0.43	3.28
p.Flamingo, 900mm high, pot 2 litre	nr	0.15	2.10	3.45	0.83	6.38
repens Alba, 900mm high, pot 0.5 litre	nr	0.10	1.40	1.15	0.38	2.93
r. rosea, 900mm high, pot 0.5 litre	nr	0.10	1.40	1.15	0.38	2.93

	Unit	Labour	Hours £	Mat'ls £	O & P £	Total £
Herbaceous plants planting (cont'd)						
Iris						
Amber Queen, 200mm high, pot 2 litre	nr	0.15	2.10	3.15	0.79	6.04
Blue Denim, 250mm high, pot 2 litre	nr	0.15	2.10	3.15	0.79	6.04
Lilli-white, 300mm high, pot 2 litre	nr	0.15	2.10	3.15	0.79	6.04
Orange Caper, 250mm high, pot 2 litre	nr	0.15	2.10	3.15	0.79	6.04
Black Swan, 1000mm high, pot 2 litre	nr	0.15	2.10	3.15	0.79	6.04
Braithwaite, 900mm high, pot 2 litre	nr	0.15	2.10	3.15	0.79	6.04
Jane Phillips, 1200mm high, pot 2 litre	nr	0.15	2.10	3.15	0.79	6.04
Rajah, 900mm high, pot 2 litre	nr	0.15	2.10	3.15	0.79	6.04
Sable, 900mm high, pot 2 litre	nr	0.15	2.10	3.15	0.79	6.04
Stepping Out, 1000mm high, pot 2 litre	nr	0.15	2.10	3.15	0.79	6.04
White Knight, 1000mm high, pot 2 litre	nr	0.15	2.10	3.15	0.79	6.04
Lupinus (Lupin)						
Chandelier, 1000mm high, pot 2 litre	nr	0.15	2.10	2.90	0.75	5.75
My Castle, 1000mm high, pot 2 litre	nr	0.15	2.10	2.90	0.75	5.75
Noble Maiden, 1000mm high, pot 2 litre	nr	0.15	2.10	2.90	0.75	5.75

	Unit	Labour	Hours £	Mat'ls £	O & P £	Total £
The Chateleine, 1000mm high, pot 2 litre	nr	0.15	2.10	2.90	0.75	5.75
The Governor, 1000mm high, pot 2 litre	nr	0.15	2.10	2.90	0.75	5.75
Paeonia (Peony)						
lactiflora, 900mm high, pot 3 litre	nr	0.15	2.10	4.90	1.05	8.05
l. Duchesse de Nemours, 1000mm high, pot 3 litre	nr	0.15	2.10	4.90	1.05	8.05
l. Festiva Maxima, 1000mm high, pot 3 litre	nr	0.15	2.10	4.90	1.05	8.05
l. Karl Rosenfeld, 1000mm high, pot 3 litre	nr	0.15	2.10	4.90	1.05	8.05
l. Sarah Bernhardt, 1000mm high, pot 3 litre	nr	0.15	2.10	4.90	1.05	8.05
officinalis Alba Plena, 750mm high, pot 3 litre	nr	0.15	2.10	4.90	1.05	8.05
o. Rosea Plena, 750mm high, pot 3 litre	nr	0.15	2.10	4.90	1.05	8.05
o. Rubra Plena, 750mm high, pot 3 litre	nr	0.15	2.10	4.90	1.05	8.05
Phlox						
douglasii Ice Mountain, 100mm high, pot 1 litre	nr	0.10	1.40	2.30	0.56	4.26

	Unit	Labour	Hours £	Mat'ls £	O & P £	Total £
Herbaceous plants planting (cont'd)						
paniculata Blue Paradise, 800mm high, pot 2 litre	nr	0.15	2.10	3.15	0.79	6.04
p. Fujiyama, 800mm high, pot 2 litre	nr	0.15	2.10	3.15	0.79	6.04
p. Jules Sandeau, 800mm high, pot 2 litre	nr	0.15	2.10	3.15	0.79	6.04
p. Kirchenfuerst, 800mm high, pot 2 litre	nr	0.15	2.10	3.15	0.79	6.04
p. Little Boy, 700mm high, pot 2 litre	nr	0.15	2.10	3.15	0.79	6.04
p. Little Laura, 600mm high, pot 2 litre	nr	0.15	2.10	3.15	0.79	6.04
p. Little Princess, 500mm high, pot 2 litre	nr	0.15	2.10	3.15	0.79	6.04
p. Mia Ruys, 800mm high, pot 2 litre	nr	0.15	2.10	3.15	0.79	6.04
p. Norah Leigh, 600mm high, pot 2 litre	nr	0.15	2.10	3.45	0.83	6.38
p. Orange Perfection, 750mm, high, pot 2 litre	nr	0.15	2.10	3.15	0.79	6.04
p. Rijnstroom, 900mm high, pot 2 litre	nr	0.15	2.10	3.15	0.79	6.04
p. Starfire, 900mm high, pot 2 litre	nr	0.15	2.10	3.15	0.79	6.04
p. White Admiral, 900mm, high, pot 0.5 litre	nr	0.10	1.40	1.15	0.38	2.93
x procumbens Variegata, 200mm, high, pot 2 litre	nr	0.15	2.10	3.15	0.79	6.04
s. Emerald Cushion Blue, 200mm, high, pot 1 litre	nr	0.10	1.40	2.30	0.56	4.26

	Unit	Labour	Hours £	Mat'ls £	O & P £	Total £
Primula						
x bulleesiana, 500mm high, pot 2 litre	nr	0.15	2.10	2.60	0.71	5.41
bulleyana, 600mm high, pot 2 litre	nr	0.15	2.10	2.60	0.71	5.41
d. Lilac, 250mm high, pot 0.5 litre	nr	0.10	1.40	1.15	0.38	2.93
d. Lilac, 250mm high, pot 2 litre	nr	0.15	2.10	2.60	0.71	5.41
d. Ruby, 250mm high, pot 0.5 litre	nr	0.10	1.40	1.15	0.38	2.93
d. Ruby, 250mm high, pot 2 litre	nr	0.15	2.10	2.60	0.71	5.41
florindae, 600mm high, pot 0.5 litre	nr	0.10	1.40	1.15	0.38	2.93
florindae, 600mm high, pot 2 litre	nr	0.15	2.10	2.60	0.71	5.41
j. Postford White, 600mm, high, pot 2 litre	nr	0.15	2.10	2.60	0.71	5.41
pulverenta, 600mm high, pot 2 litre	nr	0.15	2.10	2.60	0.71	5.41
rosae, 150mm high, pot 0.5 litre	nr	0.10	1.40	1.15	0.38	2.93
rosae, 150mm high, pot 2 litre	nr	0.15	2.10	2.90	0.75	5.75
veris, 200mm high, pot 0.5 litre	nr	0.10	1.40	1.25	0.40	3.05
vialli, 300mm high, pot 0.5 litre	nr	0.10	1.40	1.45	0.43	3.28
vulgaris acaulis, 300mm high, pot 0.5 litre	nr	0.10	1.40	1.45	0.43	3.28
vulgaris acaulis, 150mm high, pot 0.5 litre	nr	0.15	2.10	1.45	0.53	4.08

	Unit	Labour	Hours £	Mat'ls £	O & P £	Total £
Herbaceous plants planting (cont'd)						
vulgaris acaulis, 150mm high, pot 2 litre	nr	0.15	2.10	2.90	0.75	5.75
Wanda, 150mm high, pot 0.5 litre	nr	0.10	1.40	1.15	0.38	2.93
Salvia						
nemorosa Amethyst, 650mm high, pot 2 litre	nr	0.15	2.10	2.90	0.75	5.75
n. Ostfriesland, 450mm high, pot 0.5 litre	nr	0.10	1.40	1.15	0.38	2.93
n. Ostfriesland, 450mm high, pot 2 litre	nr	0.15	2.10	2.90	0.75	5.75
n. Rose Queen, 450mm high, pot 0.5 litre	nr	0.10	1.40	1.15	0.38	2.93
officinalis, 600mm high pot 1 litre	nr	0.10	1.40	2.30	0.56	4.26
o. Aurea, 300mm high pot 1 litre	nr	0.10	1.40	2.30	0.56	4.26
o. Aurea, 300mm high pot 2 litre	nr	0.10	1.40	3.45	0.73	5.58
o. Icterina, 600mm high, pot 1 litre	nr	0.10	1.40	2.30	0.56	4.26
o. Icterina, 600mm high, pot 2 litre	nr	0.15	2.10	3.45	0.83	6.38
o. Purpurascens, 600mm high, pot 1 litre	nr	0.10	1.40	2.30	0.56	4.26
o. Purpurascens, 600mm high, pot 2 litre	nr	0.15	2.10	3.45	0.83	6.38
o. Tricolor, 600mm high, pot 1 litre	nr	0.10	1.40	2.30	0.56	4.26
o. Tricolor, 600mm						

	Unit	Labour	Hours £	Mat'ls £	O & P £	Total £
high, pot 2 litre	nr	0.15	2.10	3.45	0.83	6.38
patens, 450mm high, pot 2 litre	nr	0.15	2.10	3.45	0.83	6.38
p. Cambridge Blue, 600mm high, pot 2 litre	nr	0.15	2.10	3.45	0.83	6.38
verticillata Alba, 600mm high, pot 2 litre	nr	0.15	2.10	2.90	0.75	5.75
v. Purple Rain, 450mm high, pot 2 litre	nr	0.15	2.10	2.90	0.75	5.75
Veronica						
gentianoides, 300mm high, pot 1 litre	nr	0.10	1.40	2.30	0.56	4.26
peduncularis, 150mm high, pot 1 litre	nr	0.10	1.40	2.30	0.56	4.26
prostrata, 100mm high, pot 1 litre	nr	0.10	1.40	2.30	0.56	4.26
p. Trehane, 150mm high, pot 1 litre	nr	0.10	1.40	2.30	0.56	4.26
spicata, 750mm high, pot 2 litre	nr	0.15	2.10	3.45	0.83	6.38
s. Alba, 400mm high, pot 0.5 litre	nr	0.10	1.40	1.15	0.38	2.93
Viola (Violet)						
Ardross Gem, 150mm high, pot 0.5 litre	nr	0.10	1.40	1.15	0.38	2.93
Ardross Gem, 150mm high, pot 1 litre	nr	0.10	1.40	2.30	0.56	4.26
Belmont Blue, 150mm high, pot 0.5 litre	nr	0.10	1.40	1.15	0.38	2.93
Belmont Blue, 150mm high, pot 1 litre	nr	0.10	1.40	2.30	0.56	4.26

	Unit	Labour	Hours £	Mat'ls £	O & P £	Total £
Herbaceous plants planting (cont'd)						
Buttercup, 150mm high, pot 0.5 litre	nr	0.10	1.40	1.15	0.38	2.93
Buttercup, 150mm high, pot 1 litre	nr	0.10	1.40	2.30	0.56	4.26
Dawn, 150mm high, pot 0.5 litre	nr	0.10	1.40	1.15	0.38	2.93
Dawn, 150mm high, pot 1 litre	nr	0.10	1.40	2.30	0.56	4.26
Etain, 150mm high, pot 0.5 litre	nr	0.10	1.40	1.15	0.38	2.93
Etain, 150mm high, pot 1 litre	nr	0.10	1.40	2.30	0.56	4.26
Jackanapes, 150mm high, pot 0.5 litre	nr	0.10	1.40	1.15	0.38	2.93
Jackanapes, 150mm high, pot 1 litre	nr	0.10	1.40	2.30	0.56	4.26
Janet, 150mm high, pot 0.5 litre	nr	0.10	1.40	1.15	0.38	2.93
Janet, 150mm high, pot 1 litre	nr	0.10	1.40	2.30	0.56	4.26
Labradorica, 100mm high, pot 0.5 litre	nr	0.10	1.40	1.15	0.38	2.93
Maggie Matt, 150mm high, pot 0.5 litre	nr	0.10	1.40	1.15	0.38	2.93
Maggie Matt, 150mm high, pot 1 litre	nr	0.10	1.40	2.30	0.56	4.26
Mrs Lancaster, 150mm high, pot 0.5 litre	nr	0.10	1.40	1.15	0.38	2.93
Mrs Lancaster, 150mm high, pot 1 litre	nr	0.10	1.40	2.30	0.56	4.26
odorata, 100mm high, pot 0.5 litre	nr	0.10	1.40	1.15	0.38	2.93

	Unit	Labour	Hours £	Mat'ls £	O & P £	Total £
Rebecca, 150mm high, pot 0.5 litre	nr	0.10	1.40	1.15	0.38	2.93
Rebecca, 150mm high, pot 1 litre	nr	0.10	1.40	2.30	0.56	4.26
Roscastle Black, 150mm high, pot 0.5 litre	nr	0.10	1.40	1.15	0.38	2.93
Roscastle Black, 150mm high, pot 1 litre	nr	0.10	1.40	2.30	0.56	4.26
scoria Freckles, 150mm high, pot 1 litre	nr	0.10	1.40	2.30	0.56	4.26
Zoe, 150mm high, pot 1 litre	nr	0.10	1.40	2.30	0.56	4.26

HEDGING

The following descriptions refer to dimensions and sizes of hedge plants as set out below.

Dwarf: 20 to 60cm high
Small: 60 to 120cm high
Medium: 120 to 200cm high
Large: over 200cm high

Form planting hole in cultivated area, place hedging plant in hole, backfill, water and surround with peat

Acer (Maple)

	Unit	Labour	Hours £	Mat'ls £	O & P £	Total £
campestre, small, pot 1 litre	nr	0.10	1.40	0.55	0.29	2.24
campestre, small, pot 1 litre	nr	0.10	1.40	0.60	0.30	2.30

	Unit	Labour	Hours £	Mat'ls £	O & P £	Total £
Hedging (cont'd)						
campestre, medium, pot 1 litre	nr	0.10	1.40	0.70	0.32	2.42
campestre, medium, pot 2 litre	nr	0.15	2.10	1.15	0.49	3.74
platanoides, small, pot 1 litre	nr	0.10	1.40	0.55	0.29	2.24
platanoides, small, pot 1 litre	nr	0.10	1.40	0.60	0.30	2.30
platanoised, medium, pot 1 litre	nr	0.10	1.40	0.70	0.32	2.42
platanoides, medium, pot 2 litre	nr	0.15	2.10	1.15	0.49	3.74
platanoides, small, pot 1 litre	nr	0.10	1.40	0.55	0.29	2.24
platanoides, small, pot 1 litre	nr	0.10	1.40	0.60	0.30	2.30
platanoised, medium, pot 1 litre	nr	0.10	1.40	0.70	0.32	2.42
platanoides, medium, pot 2 litre	nr	0.15	2.10	1.15	0.49	3.74
pseudoplatanus, small, pot 1 litre	nr	0.10	1.40	0.55	0.29	2.24
pseudoplatanus, small, pot 1 litre	nr	0.10	1.40	0.60	0.30	2.30
pseudoplatanus, medium, pot 1 litre	nr	0.15	2.10	0.70	0.42	3.22
pseudoplatanus, medium, pot 2 litre	nr	0.15	2.10	1.15	0.49	3.74
Alnus (Alder)						
cordata, small, pot 1 litre	nr	0.10	1.40	0.55	0.29	2.24

	Unit	Labour	Hours £	Mat'ls £	O & P £	Total £
cordata, small, pot 1 litre	nr	0.10	1.40	0.60	0.30	2.30
codarta, medium, pot 1 litre	nr	0.10	1.40	0.70	0.32	2.42
cordata, medium, pot 2 litre	nr	0.15	2.10	1.15	0.49	3.74
glutinosa, small, pot 1 litre	nr	0.10	1.40	0.55	0.29	2.24
glutinosa, small, pot 1 litre	nr	0.10	1.40	0.60	0.30	2.30
glutinosa, medium, pot 1 litre	nr	0.10	1.40	0.70	0.32	2.42
glutinosa, medium, pot 2 litre	nr	0.15	2.10	1.15	0.49	3.74
incana, small, pot 1 litre	nr	0.10	1.40	0.55	0.29	2.24
incana, small, pot 1 litre	nr	0.10	1.40	0.60	0.30	2.30
incana, medium, pot 1 litre	nr	0.10	1.40	0.70	0.32	2.42
incana, medium, pot 2 litre	nr	0.15	2.10	1.15	0.49	3.74
Betula (Birch)						
pendula, small, pot 1 litre	nr	0.10	1.40	0.55	0.29	2.24
pendula, small, pot 1 litre	nr	0.10	1.40	0.60	0.30	2.30
pendula, medium, pot 1 litre	nr	0.10	1.40	0.70	0.32	2.42
pendula, medium, pot 2 litre	nr	0.15	2.10	1.15	0.49	3.74

	Unit	Labour	Hours £	Mat'ls £	O & P £	Total £
Hedging (cont'd)						
pubescens, small, pot 1 litre	nr	0.10	1.40	0.55	0.29	2.24
pubescens, small, pot 1 litre	nr	0.10	1.40	0.60	0.30	2.30
pubescens, medium, pot 1 litre	nr	0.10	1.40	0.70	0.32	2.42
Chamaecyparis (Cypress)						
lawsoniana, small, pot 3 litre	nr	0.20	2.80	5.20	1.20	9.20
lawsoniama, small, pot 10 litre	nr	0.30	4.20	9.20	2.01	15.41
Crataegus (Hawthorn)						
laevigata, small, pot 1 litre	nr	0.10	1.40	0.55	0.29	2.24
laevigata, small, pot 1 litre	nr	0.10	1.40	0.60	0.30	2.30
laevigata, medium, pot 1 litre	nr	0.10	1.40	0.70	0.32	2.42
laevigata, medium, pot 2 litre	nr	0.15	2.10	1.15	0.49	3.74
Cupressocyparis						
leylandii, small, pot 3 litre	nr	0.20	2.80	3.45	0.94	7.19
leylandii, medium, pot 10 litre	nr	0.30	4.20	9.20	2.01	15.41
I. Castlewellan, small, pot 3 litre	nr	0.20	2.80	3.45	0.94	7.19

	Unit	Labour	Hours £	Mat'ls £	O & P £	Total £
I. Castlewellan, small, pot 3 litre	nr	0.20	2.80	3.45	0.94	7.19
Fagus (Beech)						
sylvatica, dwarf, pot 3 litre	nr	0.20	2.80	2.30	0.77	5.87
sylvatica, dwarf, pot 10 litre	nr	0.30	4.20	11.50	2.36	18.06
Fagus						
pendula, small, pot 1 litre	nr	0.10	1.40	0.70	0.32	2.42
pendula, small, pot 1 litre	nr	0.10	1.40	0.85	0.34	2.59
pendula, medium, pot 2 litre	nr	0.15	2.10	1.15	0.49	3.74
pendula, medium, pot 2 litre	nr	0.15	2.10	1.50	0.54	4.14
Atropurpurea, small, pot 1 litre	nr	0.10	1.40	1.85	0.49	3.74
Atropurpurea, small, pot 1 litre	nr	0.10	1.40	2.40	0.57	4.37
Atropurpurea, medium, pot 2 litre	nr	0.15	2.10	2.60	0.71	5.41
Atropurpurea, medium, pot 2 litre	nr	0.15	2.10	3.75	0.88	6.73
Ligustrum (Privet)						
ovalifolium, dwarf, pot 1 litre	nr	0.10	1.40	0.50	0.29	2.19
ovalifolium, small, pot 1 litre	nr	0.10	1.40	0.55	0.29	2.24

	Unit	Labour	Hours £	Mat'ls £	O & P £	Total £
Hedging (cont'd)						
ovalifolium, small, pot 1 litre	nr	0.10	1.40	0.70	0.32	2.42
ovalifolium, small, pot 3 litre	nr	0.20	2.80	2.30	0.77	5.87
vulgare, dwarf, pot 1 litre	nr	0.10	1.40	0.50	0.29	2.19
vulgare, small, pot 1 litre	nr	0.10	1.40	0.55	0.29	2.24
vulgare, small, pot 1 litre	nr	0.10	1.40	0.70	0.32	2.42
vulgare, small, pot 3 litre	nr	0.20	2.80	2.30	0.77	5.87
Prunus (Cherry)						
avium, small, pot 1 litre	nr	0.10	1.40	0.55	0.29	2.24
avium, small, pot 1 litre	nr	0.10	1.40	0.70	0.32	2.42
avium, medium, pot 1 litre	nr	0.10	1.40	1.15	0.38	2.93
cerasifera, small, pot 1 litre	nr	0.10	1.40	0.50	0.29	2.19
cerasifera, small, pot 1 litre	nr	0.10	1.40	0.55	0.29	2.24
cerasifera, medium, pot 1 litre	nr	0.10	1.40	0.70	0.32	2.42
laurocerasus, small, pot 3 litre	nr	0.20	2.80	2.30	0.77	5.87
laurocerasus, small, pot 3 litre	nr	0.20	2.80	3.45	0.94	7.19
laurocerasus, small, pot 10 litre	nr	0.30	4.20	12.65	2.53	19.38

	Unit	Labour	Hours £	Mat'ls £	O & P £	Total £
lusitanica, dwarf, pot 3 litre	nr	0.20	2.80	3.45	0.94	7.19
lusitanica, dwarf, pot 3 litre	nr	0.20	2.80	4.00	1.02	7.82
lusitanica, small, pot 10 litre	nr	0.30	4.20	12.65	2.53	19.38
padus, small, pot 1 litre	nr	0.10	1.40	0.55	0.29	2.24
padus, small, pot 1 litre	nr	0.10	1.40	0.70	0.32	2.42
padus, medium, pot 2 litre	nr	0.15	2.10	1.15	0.49	3.74
spinosa, dwarf, pot 1 litre	nr	0.10	1.40	0.50	0.29	2.19
spinosa, small, pot 1 litre	nr	0.10	1.40	0.65	0.31	2.36
spinosa, small, pot 1 litre	nr	0.10	1.40	0.95	0.35	2.70
Quercus (Oak)						
cerris, dwarf, pot 1 litre	nr	0.10	1.40	0.65	0.31	2.36
cerris, small, pot 1 litre	nr	0.10	1.40	0.85	0.34	2.59
ilex, dwarf, pot 3 litre	nr	0.20	2.80	3.45	0.94	7.19
ilex, small, pot 3 litre	nr	0.20	2.80	4.60	1.11	8.51
ilex, medium, pot 10 litre	nr	0.30	4.20	16.10	3.05	23.35
palustris, dwarf, pot 1 litre	nr	0.10	1.40	0.70	0.32	2.42
palustris, small, pot 1 litre	nr	0.10	1.40	0.85	0.34	2.59
robur, dwarf, pot 1 litre	nr	0.10	1.40	0.65	0.31	2.36
robur, small, pot 2 litre	nr	0.15	2.10	0.85	0.44	3.39

	Unit	Labour	Hours £	Mat'ls £	O & P £	Total £
Hedging (cont'd)						
robur, small, pot 1 litre	nr	0.10	1.40	0.95	0.35	2.70
robur, medium, pot 2 litre	nr	0.15	2.10	1.50	0.54	4.14
rubra, dwarf, pot 1 litre	nr	0.10	1.40	0.65	0.31	2.36
rubra, small, pot 2 litre	nr	0.15	2.10	0.85	0.44	3.39
rubra, small, pot 1 litre	nr	0.10	1.40	0.95	0.35	2.70
rubra, medium, pot 2 litre	nr	0.15	2.10	1.50	0.54	4.14
Salix (Willow)						
alba, small, pot 1litre	nr	0.10	1.40	0.55	0.29	2.24
alba, small, pot 1 litre	nr	0.10	1.40	0.70	0.32	2.42
alba, medium, pot 1 litre	nr	0.15	2.10	1.15	0.49	3.74
a. Liempde, small, pot 1 litre	nr	0.10	1.40	0.55	0.29	2.24
a. Liempde, small, pot 1 litre	nr	0.10	1.40	0.70	0.32	2.42
alba var.caerulea, small, pot 1 litre	nr	0.10	1.40	0.55	0.29	2.24
alba var.caerulea, small, pot 1 litre	nr	0.10	1.40	0.70	0.32	2.42
a. subsp. vitellina, small, pot 1 litre	nr	0.10	1.40	0.55	0.29	2.24
a. subsp. vitellina, small, pot 1 litre	nr	0.10	1.40	0.70	0.32	2.42
a. subsp. vitellina, small, pot 3 litre	nr	0.20	2.80	2.30	0.77	5.87
caprea, small, pot 1litre	nr	0.10	1.40	0.55	0.29	2.24
caprea, small, pot 1 litre	nr	0.10	1.40	0.70	0.32	2.42

	Unit	Labour	Hours	Mat'ls	O & P	Total
caprea, medium, pot 1 litre	nr	0.15	2.10	1.15	0.49	3.74
cinerea, small, pot 1litre	nr	0.10	1.40	0.55	0.29	2.24
cinerea, small, pot 1 litre	nr	0.10	1.40	0.70	0.32	2.42
cinerea, small, pot 3 litre	nr	0.20	2.80	2.30	0.77	5.87
daphnoides, small, pot 1 litre	nr	0.10	1.40	0.55	0.29	2.24
daphnoides, small, pot 1 litre	nr	0.10	1.40	0.70	0.32	2.42
daphnoides, small, pot 3 litre	nr	0.10	1.40	2.30	0.56	4.26
elaeangnos, small, pot 1 litre	nr	0.10	1.40	0.55	0.29	2.24
elaeangnos, small, pot 3 litre	nr	0.20	2.80	2.30	0.77	5.87
viminalis, small, pot 1 litre	nr	0.10	1.40	0.55	0.29	2.24
viminalis, small, pot 1 litre	nr	0.10	1.40	0.70	0.32	2.42
viminalis, small, pot 3 litre	nr	0.10	1.40	2.30	0.56	4.26
Sorbus (Whitebeam)						
aria, small, pot 1litre	nr	0.10	1.40	0.70	0.32	2.42
aria, small, pot 1litre	nr	0.10	1.40	0.90	0.35	2.65
aria, small, pot 1litre	nr	0.10	1.40	1.10	0.38	2.88
aria, medium, pot 1 litre	nr	0.15	2.10	1.50	0.54	4.14
acuparia, small, pot 1 litre	nr	0.10	1.40	0.50	0.29	2.19
acuparia, small, pot 1 litre	nr	0.10	1.40	0.60	0.30	2.30
acuparia, small, pot 1 litre	nr	0.10	1.40	0.70	0.32	2.42

	Unit	Labour	Hours £	Mat'ls £	O & P £	Total £
Hedging (cont'd)						
acuparia, medium, pot 2 litre	nr	0.15	2.10	1.15	0.49	3.74
intermedia, small, pot 1 litre	nr	0.10	1.40	0.70	0.32	2.42
intermedia, small, pot 1 litre	nr	0.10	1.40	0.90	0.35	2.65
intermedia, small, pot 1 litre	nr	0.10	1.40	1.10	0.38	2.88
intermedia, medium, pot 2 litre	nr	0.15	2.10	1.50	0.54	4.14

BEDDING PLANTS

The following items cover the work involved in the creation of a formal bedding display.The cost of supplying and bedding the plants has not been included because of the wide variation in density, design patern and quality of the plants involved.

	Unit	Labour	Hours £	Mat'ls £	O & P £	Total £
Protect edge of existing lawn with polythene sheeting (6 uses)	m	0.03	0.42	0.05	0.07	0.54
Remove existing plants from bed	m2	1.40	19.60	-	2.94	22.54

	Unit	Labour	Hours £	Mat'ls £	O & P £	Total £
Fork over to a depth of 300mm, remove weeds and rake over	m2	2.00	28.00	-	4.20	32.20
Cut edge of existing lawn with shears	m	0.03	0.42	-	0.06	0.48
Dig trench 300mm deep and fork in well-rotted manure 100mm deep to bottom of trench	m	0.12	1.68	0.15	0.27	2.10

MAINTENANCE WORK BY HAND

Grassed areas

	Unit	Labour	Hours £	Mat'ls £	O & P £	Total £
Pick up litter and remove	100m2	0.15	2.10	1.50	0.54	4.14
Rake up loose grass and place in heaps for disposal	100m2	0.35	4.90	1.50	0.96	7.36
Trim edges of grassed areas with edging tool	100m	0.50	7.00	1.50	1.28	9.78
Aerate by forking	100m2	1.50	21.00	1.50	3.38	25.88
Sweep up corings and place in heaps for disposal	100m2	1.50	21.00	1.50	3.38	25.88
Apply weedkiller (175 kg per ha)	100m2	0.50	7.00	0.30	1.10	8.40

	Unit	Labour	Hours £	Mat'ls £	O & P £	Total £
Apply fertiliser (35g per) ha)	100m2	0.50	7.00	0.30	1.10	8.40
Planted areas						
Remove stones and debris and place in heaps for disposal	100m2	1.00	14.00	0.30	2.15	16.45
Weed and hoe and place weeds in heaps for disposal	100m2	1.00	14.00	0.30	2.15	16.45
Fine-grade bark mulch, thickness						
50mm	100m2	1.50	21.00	130.00	22.65	173.65
75mm	100m2	1.75	24.50	195.00	32.93	252.43
100mm	100m2	2.00	28.00	260.00	43.20	331.20
Medium-grade bark mulch, thickness						
50mm	100m2	1.50	21.00	188.00	31.35	240.35
75mm	100m2	1.75	24.50	282.00	45.98	352.48
100mm	100m2	2.00	28.00	366.00	59.10	453.10
Coarse-grade bark mulch, thickness						
50mm	100m2	1.50	21.00	224.00	36.75	281.75
75mm	100m2	1.75	24.50	336.00	54.08	414.58
100mm	100m2	2.00	28.00	448.00	71.40	547.40

	Unit	Plant £	Mat'ls £	O & P £	Total £
MAINTENANCE WORK BY MACHINE					
Where applicable the plant column includes the cost of the operator.					
Grassed areas					
Sweep up leaves with motorised vacuum cleaner	100m2	0.84	-	0.13	0.97
Cut grass to specified height with					
petrol-powered mower	100m2	2.40	-	0.36	2.76
multi-unit mower	100m2	0.48	-	0.07	0.55
ride-on rotary mower	100m2	0.67	-	0.10	0.77
Aerate grassed surfaces with					
slitter aerator	100m2	5.56	-	0.83	6.39
tractor-drawn aeratot	100m2	3.22	-	0.48	3.70
Harrow grassed surfaces with					
chain harrow	100m2	0.98	-	0.15	1.13
drag mat	100m2	1.10	-	0.17	1.27

	Unit	Labour	Hours £	Mat'ls £	O & P £	Total £
SUNDRIES						
Treated softwood tree stakes driven into the ground						
50mm diameter						
1.7m long	nr	0.10	1.40	1.88	0.49	3.77
2.0m long	nr	0.12	1.68	2.20	0.58	4.46
2.4m long	nr	0.14	1.96	2.65	0.69	5.30
100mm diameter						
1.7m long	nr	0.14	1.96	2.48	0.67	5.11
2.0m long	nr	0.16	2.24	2.98	0.78	6.00
2.4m long	nr	0.18	2.52	3.44	0.89	6.85
Cleft chestnut tree stakes driven into the ground						
50-70mm wide						
1.7m long	nr	0.10	1.40	1.63	0.45	3.49
2.0m long	nr	0.12	1.68	1.98	0.55	4.21
2.4m long	nr	0.14	1.96	2.37	0.65	4.98
Tree ties, buckle type						
25 x 450mm	nr	0.05	0.70	0.45	0.17	1.32
25 x 600mm	nr	0.05	0.70	0.55	0.19	1.44
Galvanised wire net tree guards						
150mm diameter x 450mm high	nr	0.30	4.20	7.55	1.76	13.51
150mm diameter x 600mm high	nr	0.30	4.20	8.63	1.92	14.75

		Unit	Plant £	Mat'ls £	O & P £	Total £
200mm diameter x 1000mm high	nr	0.35	4.90	8.88	2.07	15.85
200mm diameter x 1200mm high	nr	0.35	4.90	9.34	2.14	16.38
250mm diameter x 1400mm high	nr	0.35	4.90	10.56	2.32	17.78
250mm diameter x 1800mm high	nr	0.35	4.90	11.89	2.52	19.31
Bio-degradable tree guards						
600mm high	nr	0.10	1.40	0.98	0.36	2.74
750mm high	nr	0.10	1.40	1.15	0.38	2.93
900mm high	nr	0.10	1.40	1.32	0.41	3.13
1200mm high	nr	0.10	1.40	1.48	0.43	3.31

Part Two

UNIT RATES

Hard landscaping

Brick walling

Masonry

Damp-proof course

Precast concrete

Kerbs and edgings

Sub-bases

Beds and pavings

Fencing

Drainage

BRICK WALLING

Common bricks (basic price £140 per thousand) in cement mortar (1:3), stretcher bond

	Unit	Labour	Hours £	Mat'ls £	O & P £	Total £
walls						
half brick thick	m2	1.70	23.80	11.45	5.29	40.54
half brick thick curved	m2	2.30	32.20	11.45	6.55	50.20
one brick thick	m2	2.80	39.20	22.90	9.32	71.42
one brick thick curved	m2	3.40	47.60	22.90	10.58	81.08
one and a half brick thick	m2	3.50	49.00	34.35	12.50	95.85
two brick thick	m2	4.20	58.80	45.80	15.69	120.29
two brick thick battered	m2	4.80	67.20	45.80	16.95	129.95
walls, facework one side						
half brick thick	m2	1.80	25.20	11.45	5.50	42.15
half brick thick curved	m2	2.40	33.60	11.45	6.76	51.81
one brick thick	m2	2.90	40.60	22.90	9.53	73.03
one brick thick curved	m2	3.50	49.00	22.90	10.79	82.69
one and a half brick thick	m2	3.60	50.40	34.35	12.71	97.46
two brick thick	m2	4.30	60.20	45.80	15.90	121.90
two brick thick battered	m2	4.90	68.60	45.80	17.16	131.56
walls, facework both sides						
half brick thick	m2	1.90	26.60	11.45	5.71	43.76
half brick thick curved	m2	2.50	35.00	11.45	6.97	53.42
one brick thick		3.00	42.00	22.90	9.74	74.64
one brick thick curved	m2	3.60	50.40	22.90	11.00	84.30
one and a half brick thick	m2	3.70	51.80	22.90	11.21	85.91

	Unit	Labour	Hours £	Mat'ls £	O & P £	Total £
Common bricks (cont'd)						
two brick thick	m2	4.40	61.60	45.80	16.11	123.51
two brick thick battered	m2	5.00	70.00	45.80	17.37	133.17
Common bricks (basic price £200 per thousand) in cement mortar (1:3), stretcher bond						
walls						
half brick thick	m2	1.70	23.80	16.45	6.04	46.29
half brick thick curved	m2	2.30	32.20	16.45	7.30	55.95
one brick thick	m2	2.80	39.20	32.90	10.82	82.92
one brick thick curved	m2	3.40	47.60	32.90	12.08	92.58
one and a half brick thick	m2	3.50	49.00	49.35	14.75	113.10
two brick thick	m2	4.20	58.80	49.35	16.22	124.37
two brick thick battered	m2	4.80	67.20	65.80	19.95	152.95
walls, facework one side						
half brick thick	m2	1.80	25.20	16.45	6.25	47.90
half brick thick curved	m2	2.40	33.60	16.45	7.51	57.56
one brick thick	m2	2.90	40.60	32.90	11.03	84.53
one brick thick curved	m2	3.50	49.00	32.90	12.29	94.19
one and a half brick thick	m2	3.60	50.40	49.35	14.96	114.71
two brick thick	m2	4.30	60.20	49.35	16.43	125.98
two brick thick battered	m2	4.90	68.60	65.80	20.16	154.56
walls, facework both sides						
half brick thick	m2	1.90	26.60	16.45	6.46	49.51

	Unit	Labour	Hours £	Mat'ls £	O & P £	Total £
half brick thick curved	m2	2.50	35.00	16.45	7.72	59.17
one brick thick	m2	3.00	42.00	32.90	11.24	86.14
one brick thick curved	m2	3.60	50.40	32.90	12.50	95.80
one and a half brick thick	m2	3.70	62.53	49.80	16.85	129.18
two brick thick	m2	4.40	74.36	86.40	24.11	184.87
two brick thick battered	m2	5.00	70.00	65.80	20.37	156.17
Facing bricks (basic price £250 per thousand) in cement mortar (1:3), English garden wall bond						
walls, facework one side						
half brick thick	m2	1.80	25.20	23.41	7.29	55.90
half brick thick curved	m2	2.40	33.60	23.41	8.55	65.56
one brick thick	m2	2.90	40.60	62.38	15.45	118.43
one brick thick curved	m2	3.50	49.00	62.38	16.71	128.09
one and a half brick thick	m2	3.60	50.40	76.68	19.06	146.14
two brick thick	m2	4.30	60.20	102.38	24.39	186.97
two brick thick battered	m2	4.90	68.60	102.38	25.65	196.63
walls, facework both sides						
half brick thick	m2	1.90	26.60	23.41	7.50	57.51
half brick thick curved	m2	2.50	35.00	23.41	8.76	67.17
one brick thick	m2	3.00	42.00	62.38	15.66	120.04
one brick thick curved	m2	3.60	50.40	62.38	16.92	129.70
one and a half brick thick	m2	3.70	51.80	76.68	19.27	147.75
two brick thick	m2	4.40	61.60	102.38	24.60	188.58
two brick thick battered	m2	5.00	70.00	102.38	25.86	198.24

	Unit	Labour	Hours £	Mat'ls £	O & P £	Total £
Isolated casings, half brick thick, faced all round	m2	2.30	32.20	23.41	8.34	63.95
Isolated piers, one brick thick, faced all round	m2	2.30	32.20	62.38	14.19	108.77
Arches						
215mm high, 102mm wide	m2	1.30	18.20	4.36	3.38	25.94
215mm high, 215mm wide	m2	2.00	28.00	7.34	5.30	40.64
215mm high, 102mm wide, segmental	m2	2.70	37.80	8.30	6.92	53.02
215mm high, 315mm wide, segmental	m2	3.80	53.20	12.00	9.78	74.98
Facing bricks (basic price £350 per thousand) in cement mortar (1:3), English garden wall bond						
walls, facework one side						
half brick thick	m2	1.80	25.20	28.63	8.07	61.90
half brick thick curved	m2	2.40	33.60	28.63	9.33	71.56
one brick thick	m2	2.90	40.60	57.26	14.68	112.54
one brick thick curved	m2	3.50	49.00	57.26	15.94	122.20
one and a half brick thick	m2	3.60	50.40	85.89	20.44	156.73
two brick thick	m2	4.30	60.20	114.57	26.22	200.99
two brick thick battered	m2	4.90	68.60	114.57	27.48	210.65
walls, facework both sides						
half brick thick	m2	1.90	32.11	28.63	9.11	69.85
half brick thick curved	m2	2.50	40.56	28.63	10.38	79.57
one brick thick	m2	3.00	50.70	57.26	16.19	124.15

	Unit	Hours	Hours £	Mat'ls £	O & P £	Total £
one brick thick curved	m2	3.60	60.84	57.26	17.72	135.82
one and a half brick thick	m2	3.70	62.53	85.89	22.26	170.68
two brick thick	m2	4.40	74.36	114.57	28.34	217.27
two brick thick battered	m2	5.00	70.00	114.57	27.69	212.26
Isolated casings, half brick thick, faced all round	m2	2.30	32.20	28.63	9.12	69.95
Isolated piers, one brick thick, faced all round	m2	2.30	32.20	57.26	13.42	102.88
Arches						
215mm high, 102mm wide	m2	1.30	18.20	6.30	3.68	28.18
215mm high, 215mm wide	m2	2.00	28.00	6.64	5.20	39.84
215mm high, 102mm wide, segmental	m2	2.70	37.80	7.52	6.80	52.12
215mm high, 315mm wide, segmental	m2	3.80	53.20	10.80	9.60	73.60
Facing bricks (basic price £500 per thousand) in cement mortar (1:3), English garden wall bond						
walls, facework one side						
half brick thick	m2	1.80	25.20	34.85	9.01	69.06
half brick thick curved	m2	2.40	33.60	34.85	10.27	78.72
one brick thick	m2	2.90	40.60	69.70	16.55	126.85
one brick thick curved	m2	3.50	49.00	69.70	17.81	136.51
one and a half brick thick	m2	3.60	50.40	103.55	23.09	177.04

	Unit	Labour	Hours £	Mat'ls £	O & P £	Total £
Facings £500 per thousand (cont'd)						
two brick thick	m2	4.30	60.20	139.40	29.94	229.54
two brick thick battered	m2	4.90	68.60	139.40	31.20	239.20
walls, facework both sides						
half brick thick	m2	1.90	32.11	34.85	10.04	77.00
half brick thick curved	m2	2.50	40.56	34.85	11.31	86.72
one brick thick	m2	3.00	50.70	69.70	18.06	138.46
one brick thick curved	m2	3.60	60.84	69.70	19.58	150.12
one and a half brick thick	m2	3.70	62.53	103.55	24.91	190.99
two brick thick	m2	4.40	74.36	139.40	32.06	245.82
two brick thick battered	m2	5.00	70.00	139.40	31.41	240.81
Isolated casings, half brick thick, faced all round	m2	2.30	32.20	34.85	10.06	77.11
Isolated piers, one brick thick, faced all round	m2	2.30	32.20	34.85	10.06	77.11
Arches						
215mm high, 102mm wide	m2	1.30	18.20	8.73	4.04	30.97
215mm high, 215mm wide	m2	2.00	28.00	14.68	6.40	49.08
215mm high, 102mm wide, segmental	m2	2.70	37.80	16.54	8.15	62.49
215mm high, 315mm wide, segmental	m2	3.80	53.20	23.13	11.45	87.78

	Unit	Labour	Hours £	Mat'ls £	O & P £	Total £
Facing bricks (basic price £250 per thousand) in cement mortar (1:3), flemish bond						
walls, facework one side						
half brick thick	m2	1.80	25.20	23.41	7.29	55.90
half brick thick curved	m2	2.40	33.60	23.41	8.55	65.56
one brick thick	m2	2.90	40.60	62.38	15.45	118.43
one brick thick curved	m2	3.50	49.00	62.38	16.71	128.09
one and a half brick thick	m2	3.60	50.40	76.68	19.06	146.14
two brick thick	m2	4.30	60.20	102.38	24.39	186.97
two brick thick battered	m2	4.90	68.60	102.38	25.65	196.63
walls, facework both sides						
half brick thick	m2	1.90	32.11	23.41	8.33	63.85
half brick thick curved	m2	2.50	40.56	23.41	9.60	73.57
one brick thick	m2	3.00	50.70	62.38	16.96	130.04
one brick thick curved	m2	3.60	60.84	62.38	18.48	141.70
one and a half brick thick	m2	3.70	62.53	76.68	20.88	160.09
two brick thick	m2	4.40	74.36	102.38	26.51	203.25
two brick thick battered	m2	5.00	70.00	102.38	25.86	198.24
Isolated casings, half brick thick, faced all round	m2	2.30	32.20	23.41	8.34	63.95
Isolated piers, one brick thick, faced all round	m2	2.30	32.20	62.38	14.19	108.77

	Unit	Labour	Hours £	Mat'ls £	O & P £	Total £
Arches						
215mm high, 102mm wide	m2	1.30	18.20	4.36	3.38	25.94
215mm high, 215mm wide	m2	2.00	28.00	7.34	5.30	40.64
215mm high, 102mm wide, segmental	m2	2.70	37.80	8.30	6.92	53.02
215mm high, 315mm wide, segmental	m2	3.80	53.20	12.00	9.78	74.98
Facing bricks (basic price £350 per thousand) in cement mortar (1:3), flemish bond						
walls, facework one side						
half brick thick	m2	1.80	25.20	28.63	8.07	61.90
half brick thick curved	m2	2.40	33.60	28.63	9.33	71.56
one brick thick	m2	2.90	40.60	57.26	14.68	112.54
one brick thick curved	m2	3.50	49.00	57.26	15.94	122.20
one and a half brick thick	m2	3.60	50.40	85.89	20.44	156.73
two brick thick	m2	4.30	60.20	114.57	26.22	200.99
two brick thick battered	m2	4.90	68.60	114.57	27.48	210.65
walls, facework both sides						
half brick thick	m2	1.90	32.11	28.63	9.11	69.85
half brick thick curved	m2	2.50	40.56	28.63	10.38	79.57
one brick thick	m2	3.00	50.70	57.26	16.19	124.15
one brick thick curved	m2	3.60	60.84	57.26	17.72	135.82
one and a half brick thick	m2	3.70	62.53	85.89	22.26	170.68
two brick thick	m2	4.40	74.36	114.57	28.34	217.27
two brick thick battered	m2	5.00	70.00	114.57	27.69	212.26

	Unit	Labour	Hours £	Mat'ls £	O & P £	Total £
Isolated casings, half brick thick, faced all round	m2	2.30	32.20	28.63	9.12	69.95
Isolated piers, one brick thick, faced all round	m2	2.30	32.20	57.26	13.42	102.88
Arches						
215mm high, 102mm wide	m2	1.30	18.20	6.30	3.68	28.18
215mm high, 215mm wide	m2	2.00	28.00	6.64	5.20	39.84
215mm high, 102mm wide, segmental	m2	2.70	37.80	7.52	6.80	52.12
215mm high, 315mm wide, segmental	m2	3.80	53.20	10.80	9.60	73.60
Facing bricks (basic price £500 per thousand) in cement mortar (1:3), flemish bond						
walls, facework one side						
half brick thick	m2	1.80	25.20	34.85	9.01	69.06
half brick thick curved	m2	2.40	33.60	34.85	10.27	78.72
one brick thick	m2	2.90	40.60	69.70	16.55	126.85
one brick thick curved	m2	3.50	49.00	69.70	17.81	136.51
one and a half brick thick	m2	3.60	50.40	103.55	23.09	177.04
two brick thick	m2	4.30	60.20	139.40	29.94	229.54
two brick thick battered	m2	4.90	68.60	139.40	31.20	239.20
walls, facework both sides						
half brick thick	m2	1.90	32.11	34.85	10.04	77.00
half brick thick curved	m2	2.50	40.56	34.85	11.31	86.72
one brick thick	m2	3.00	50.70	69.70	18.06	138.46

	Unit	Labour	Hours £	Mat'ls £	O & P £	Total £
Facings £500 per thousand (cont'd)						
one brick thick curved	m2	3.60	60.84	69.70	19.58	150.12
one and a half brick thick	m2	3.70	62.53	103.55	24.91	190.99
two brick thick	m2	4.40	74.36	139.40	32.06	245.82
two brick thick battered	m2	5.00	70.00	139.40	31.41	240.81
Isolated casings, half brick thick, faced all round	m2	2.30	32.20	34.85	10.06	77.11
Isolated piers, one brick thick, faced all round	m2	2.30	32.20	34.85	10.06	77.11
Arches						
215mm high, 102mm wide	m2	1.30	18.20	8.73	4.04	30.97
215mm high, 215mm wide	m2	2.00	28.00	14.68	6.40	49.08
215mm high, 102mm wide, segmental	m2	2.70	37.80	16.54	8.15	62.49
215mm high, 315mm wide, segmental	m2	3.80	53.20	23.13	11.45	87.78
MASONRY						
Random rubble walling, laid dry, thickness						
300mm	m2	3.00	42.00	50.38	13.86	106.24
450mm	m2	3.50	49.00	75.52	18.68	143.20
500mm	m2	3.75	52.50	85.20	20.66	158.36

	Unit	Labour	Hours £	Mat'ls £	O & P £	Total £
Random rubble walling, laid dry, battered one side, thickness						
300mm	m2	3.25	45.50	50.38	14.38	110.26
450mm	m2	3.75	52.50	75.52	19.20	147.22
500mm	m2	4.00	56.00	85.20	21.18	162.38
Random rubble walling, laid dry, battered both sides, thickness						
300mm	m2	3.50	49.00	50.38	14.91	114.29
450mm	m2	4.00	56.00	75.52	19.73	151.25
500mm	m2	4.25	59.50	85.20	21.71	166.41
Random rubble walling, laid in gauged mortar (1:1:6), thickness						
300mm	m2	3.20	44.80	54.15	14.84	113.79
450mm	m2	3.70	51.80	81.18	19.95	152.93
500mm	m2	3.90	54.60	91.20	21.87	167.67
Random rubble walling, laid in gauged mortar (1:1:6), thickness, battered one side						
300mm	m2	3.45	48.30	54.15	15.37	117.82
450mm	m2	4.00	56.00	81.18	20.58	157.76
500mm	m2	4.15	58.10	91.20	22.40	171.70

	Unit	Labour	Hours £	Mat'ls £	O & P £	Total £
Rubble walling (cont'd)						
Random rubble walling, laid in gauged mortar (1:1:6), thickness, battered both sides						
300mm	m2	3.95	55.30	54.15	16.42	125.87
450mm	m2	4.50	63.00	81.18	21.63	165.81
500mm	m2	4.65	65.10	91.20	23.45	179.75
Irregular coursed rubble walling, laid in gauged mortar (1:1:6), thickness						
300mm	m2	2.10	29.40	54.15	12.53	96.08
450mm	m2	2.90	40.60	81.18	18.27	140.05
500mm	m2	3.40	47.60	91.20	20.82	159.62
Coursed rubble walling, laid in gauged mortar (1:1:6), thickness						
300mm	m2	2.40	33.60	54.15	13.16	100.91
450mm	m2	3.20	44.80	81.18	18.90	144.88
500mm	m2	3.70	51.80	91.20	21.45	164.45
Fair raking cutting on stone walling, thickness						
300mm	m	0.80	11.20	2.88	2.11	16.19
450mm	m	1.20	16.80	4.66	3.22	24.68
500mm	m	1.80	25.20	5.36	4.58	35.14

	Unit	Labour	Hours £	Mat'ls £	O & P £	Total £
Form level bed on stone walling, thickness						
300mm	m	0.20	2.80	0.82	0.54	4.16
450mm	m	0.35	4.90	1.32	0.93	7.15
500mm	m	0.50	7.00	1.44	1.27	9.71
Sundries						
Form holes for pipes up to 50mm diameter through stone walling, thickness						
300mm	m	1.20	16.80	0.00	2.52	19.32
450mm	m	1.80	25.20	0.00	3.78	28.98
500mm	m	2.65	37.10	0.00	5.57	42.67
Form holes for pipes 50-100mm diameter through stone walling, thickness						
300mm	m	1.40	19.60	0.00	2.94	22.54
450mm	m	2.00	28.00	0.00	4.20	32.20
500mm	m	2.85	39.90	0.00	5.99	45.89
Form holes for pipes over 100mm diameter through stone walling, thickness						
300mm	m	2.40	33.60	0.00	5.04	38.64
450mm	m	3.00	42.00	0.00	6.30	48.30
500mm	m	3.85	53.90	0.00	8.09	61.99

	Unit	Labour	Hours £	Mat'ls £	O & P £	Total £
Rubble walling (cont'd)						
Build in ends of steel sections in stone walling, size						
not exceeding 250mm deep	m	0.70	9.80	0.00	1.47	11.27
250-500mm deep	m	0.90	12.60	0.00	1.89	14.49
over 500mm deep	m	1.10	15.40	0.00	2.31	17.71
Mortices in stone walling, size						
50 × 50 × 100mm	nr	0.60	8.40	0.00	1.26	9.66
50 × 50 × 150mm	nr	0.65	9.10	0.00	1.37	10.47
75 × 75 × 100mm	nr	0.70	9.80	0.00	1.47	11.27
75 × 75 × 150mm	nr	0.75	10.50	0.00	1.58	12.08
Grout up mortices in cement mortar (1:3)						
50 × 50 × 100mm	nr	0.10	1.40	0.01	0.21	1.62
50 × 50 × 150mm	nr	0.15	2.10	0.18	0.34	2.62
75 × 75 × 100mm	nr	0.15	2.10	0.30	0.36	2.76
75 × 75 × 150mm	nr	0.20	2.80	0.35	0.47	3.62
Damp proof courses						
Hessian based bitumen damp-proof course bedded in gauged mortar(1:1:6)						
horizontal, width						
over 225mm	m2	0.35	4.90	4.22	1.37	10.49
112mm	m	0.05	0.70	0.56	0.19	1.45

	Unit	Labour	Hours £	Mat'ls £	O & P £	Total £
vertical, width						
over 225mm	m2	0.40	5.60	4.22	1.47	11.29
112mm	m	0.07	0.98	0.56	0.23	1.77
Fibre-based bitumen damp-proof course bedded in gauged mortar(1:1:6)						
horizontal, width						
over 225mm	m2	0.40	5.60	5.56	1.67	12.83
112mm	m	0.05	0.70	0.76	0.22	1.68
vertical, width						
over 225mm	m2	0.45	6.30	5.56	1.78	13.64
112mm	m	0.07	0.98	0.75	0.26	1.99
Pitch polymer damp proof course in gauged mortar (1:1:6)						
horizontal, width						
over 225mm	m2	0.40	5.60	7.62	1.98	15.20
112mm	m	0.05	0.70	0.90	0.24	1.84
vertical, width						
over 225mm	m2	0.45	6.30	7.62	2.09	16.01
112mm	m	0.07	0.98	0.90	0.28	2.16
Bitumen damp proof course with lead core in mortar (1:1:6)						
horizontal, width						
over 225mm	m2	0.50	7.00	16.25	3.49	26.74
112mm	m	0.06	0.84	2.37	0.48	3.69

	Unit	Labour	Hours £	Mat'ls £	O & P £	Total £
Damp-proof courses (cont'd)						
vertical, width						
over 225mm	m2	0.55	7.70	16.25	3.59	27.54
112mm	m	0.08	1.12	2.37	0.52	4.01
Two courses of slates bedded in cement mortar (1:3)						
horizontal, width						
over 225mm	m2	0.87	12.18	15.86	4.21	32.25
112mm	m	0.33	4.62	2.11	1.01	7.74
vertical, width						
over 225mm	m2	0.95	13.30	15.86	4.37	33.53
112mm	m	0.35	4.90	2.11	1.05	8.06
PRECAST CONCRETE						
Precast concrete copings twice weatherted and twice throated, size						
75 × 150 × 900mm	nr	0.30	4.20	6.84	1.66	12.70
75 × 300 × 900mm	nr	0.40	5.60	6.84	1.87	14.31
Precast concrete pier caps, four times weathered, size						
300 × 300 × 75mm	nr	1.00	14.00	6.58	3.09	23.67
450 × 450 × 75mm	nr	1.20	16.80	9.79	3.99	30.58

	Unit	Labour	Hours £	Mat'ls £	O & P £	Total £
KERBS AND EDGINGS						
Excavate trench by hand for kerb foundation						
200 × 75mm	m	0.10	1.40	0.00	0.21	1.61
250 × 100mm	m	0.12	1.68	0.00	0.25	1.93
300 × 100mm	m	0.14	1.96	0.00	0.29	2.25
450 × 150mm	m	0.20	2.80	0.00	0.42	3.22
600 × 200mm	m	0.40	5.60	0.00	0.84	6.44
Excavate curved trench by hand for kerb foundation						
250 × 100mm	m	0.12	1.68	0.00	0.25	1.93
250 × 100mm	m	0.14	1.96	0.00	0.29	2.25
300 × 100mm	m	0.16	2.24	0.00	0.34	2.58
600 × 200mm	m	0.24	3.36	0.00	0.50	3.86
600 × 200mm	m	0.44	6.16	0.00	0.92	7.08
Site-mixed concrete in foundation for kerb size						
200 × 75mm	m	0.03	0.42	1.10	0.23	1.75
250 × 100mm	m	0.04	0.56	1.80	0.35	2.71
300 × 100mm	m	0.05	0.70	2.16	0.43	3.29
450 × 150mm	m	0.18	2.52	4.70	1.08	8.30
600 × 200mm	m	0.30	4.20	8.68	1.93	14.81
Precast concrete kerbs, channels and edgings, jointed and pointed in cement mortar						
kerbs, straight						
127 × 254mm	m	0.40	5.60	6.12	1.76	13.48
152 × 305mm	m	0.45	6.30	7.68	2.10	16.08

	Unit	Labour	Hours £	Mat'ls £	O & P £	Total £
Kerbs and edgings (cont'd)						
kerbs, curved						
127 × 254mm	m	0.50	7.00	7.38	2.16	16.54
152 × 305mm	m	0.55	7.70	11.58	2.89	22.17
channels, straight						
127 × 254mm	m	0.40	5.60	7.23	1.92	14.75
channels, curved						
127 × 254mm	m	0.50	7.00	8.31	2.30	17.61
edgings, straight						
51 × 152mm	m	0.30	4.20	3.15	1.10	8.45
51 × 203mm	m	0.30	4.20	3.68	1.18	9.06
SUB-BASES						
Beds and bases compacting in layers and grading						
average thickness, 100mm						
granular fill	m2	0.10	1.40	3.02	0.66	5.08
sand	m2	0.12	1.68	2.80	0.67	5.15
hardcore	m2	0.13	1.82	2.68	0.68	5.18
average thickness, 150mm						
granular fill	m2	0.12	1.68	3.77	0.82	6.27
sand	m2	0.14	1.96	3.51	0.82	6.29
hardcore	m2	0.16	2.24	3.43	0.85	6.52
average thickness, 200mm						
granular fill	m2	0.14	1.96	4.90	1.03	7.89
sand	m2	0.16	2.24	4.43	1.00	7.67
hardcore	m2	0.18	2.52	4.25	1.02	7.79

	Unit	Labour	Hours £	Mat'ls £	O & P £	Total £
BEDS AND PAVINGS						
Preparatory work						
Excavate by hand to form path, depth						
100mm	m2	0.25	3.50	0.00	0.53	4.03
150mm	m2	0.35	4.90	0.00	0.74	5.64
200mm	m2	0.45	6.30	0.00	0.95	7.25
250mm	m2	0.60	8.40	0.00	1.26	9.66
300mm	m2	0.75	10.50	0.00	1.58	12.08
Hardcore bed compacted in layers and blinded with sand						
100mm	m2	0.08	1.12	2.80	0.59	4.51
150mm	m2	0.10	1.40	4.40	0.87	6.67
200mm	m2	0.12	1.68	5.58	1.09	8.35
Concrete beds						
Site-mixed concrete in beds						
not exceeding 150mm thick	m3	2.80	39.20	80.96	18.02	138.18
150 to 450mm thick	m3	2.20	30.80	80.96	16.76	128.52
Formwork to sides of concrete bases						
not exceeding 250mm wide	m	0.65	9.10	1.87	1.65	12.62
250 to 500mm wide	m	0.90	12.60	3.07	2.35	18.02

	Unit	Labour	Hours £	Mat'ls £	O & P £	Total £
Concrete beds (cont'd)						
Steel fabric reinforcement laid in concrete beds						
ref A142, 2.22kg/m2	m2	0.15	2.10	1.44	0.53	4.07
ref A193, 3.02kg/m2	m2	0.18	2.52	1.73	0.64	4.89
Expansion joint, impregnated fibre-base joint filler, formed joint						
12.5mm thick						
not exceeding 150mm \	m	0.18	2.52	2.34	0.73	5.59
150 to 300mm wide	m	0.24	3.36	3.75	1.07	8.18
300 to 450mm wide	m	0.30	4.20	5.67	1.48	11.35
25mm thick						
not exceeding 150mm \	m	0.20	2.80	3.33	0.92	7.05
150 to 300mm wide	m	0.26	3.64	4.96	1.29	9.89
300 to 450mm wide	m	0.32	4.48	7.71	1.83	14.02
Treat surfaces of concrete before setting						
tamping	m2	0.06	0.84	0.00	0.13	0.97
floating	m2	0.10	1.40	0.00	0.21	1.61
trowelling	m2	0.15	2.10	0.00	0.32	2.42
Granular pavings						
Washed pit gravel, 19mm, bed thickness						
50mm	m2	0.06	0.84	2.14	0.45	3.43
75mm	m2	0.08	1.12	3.21	0.65	4.98

	Unit	Labour	Hours £	Mat'ls £	O & P £	Total £
Washed pit gravel, 38mm, bed thickness						
50mm	m2	0.06	0.84	1.98	0.42	3.24
75mm	m2	0.08	1.12	2.87	0.60	4.59
Hoggin bed laid on hard-core, bed thickness						
100mm	m2	0.14	1.96	2.34	0.65	4.95
150mm	m2	0.18	2.52	3.48	0.90	6.90
200mm	m2	0.22	3.08	4.31	1.11	8.50
Bark laid on prepared surfaces, bed thickness 100mm	m2	0.05	0.70	2.14	0.43	3.27
Impregnated softwood edging boards with 38 × 38 × 225mm pegs at 1200mm centres						
38 × 100mm	m2	0.05	0.70	2.02	0.41	3.13
38 × 150mm	m2	0.07	0.98	2.32	0.50	3.80

Brick/block pavings

	Unit	Labour	Hours £	Mat'ls £	O & P £	Total £
Brick paving, £250 per 1000, 215 × 103 × 65mm, laid to falls and cross falls bedding in cement mortar 15mm thick						
straight joints both ways						
bricks laid flat	m2	1.00	14.00	23.79	5.67	43.46
bricks laid on edge	m2	1.25	17.50	36.25	8.06	61.81

	Unit	Labour	Hours £	Mat'ls £	O & P £	Total £
herringbone pattern						
bricks laid flat	m2	1.25	17.50	23.79	6.19	47.48
bricks laid on edge	m2	1.50	21.00	36.25	8.59	65.84
Brick paving, £350 per 1000, 215 × 103 × 65mm, laid to falls and cross falls bedding in cement mortar 15mm thick						
straight joints both ways						
bricks laid flat	m2	1.00	14.00	28.63	6.39	49.02
bricks laid on edge	m2	1.25	17.50	41.23	8.81	67.54
herringbone pattern						
bricks laid flat	m2	1.25	17.50	28.63	6.92	53.05
bricks laid on edge	m2	1.50	21.00	41.23	9.33	71.56
Brick paving, £500 per 1000, 215 × 103 × 65mm, laid to falls and cross falls bedding in cement mortar 15mm thick						
straight joints both ways						
bricks laid flat	m2	1.00	14.00	34.85	7.33	56.18
bricks laid on edge	m2	1.25	17.50	47.55	9.76	74.81
herringbone pattern						
bricks laid flat	m2	1.25	17.50	34.85	7.85	60.20
bricks laid on edge	m2	1.50	21.00	47.55	10.28	78.83

	Unit	Labour	Hours £	Mat'ls £	O & P £	Total £
Precast concrete pavings						
Precast concrete paving flags, spot bedded in cement mortar, straight both ways, jointing in cement mortar brushed in						
natural colour						
450 × 450 × 50mm	m2	0.55	7.70	15.02	3.41	26.13
450 × 600 × 50mm	m2	0.55	7.70	12.45	3.02	23.17
600 × 600 × 50mm	m2	0.50	7.00	10.22	2.58	19.80
600 × 900 × 50mm	m2	0.45	6.30	9.49	2.37	18.16
coloured						
450 × 450 × 50mm	m2	0.55	7.70	18.64	3.95	30.29
450 × 600 × 50mm	m2	0.55	7.70	15.67	3.51	26.88
600 × 600 × 50mm	m2	0.50	7.00	14.37	3.21	24.58
600 × 900 × 50mm	m2	0.45	6.30	12.83	2.87	22.00
Reconstituted York stone paving flags, spot bedded in cement mortar, straight both ways, jointing in cement mortar brushed in, 45mm thick, size						
300 x 300mm	m2	1.00	14.00	26.31	6.05	46.36
300 x 450mm	m2	0.80	11.20	25.44	5.50	42.14
300 x 600mm	m2	0.60	8.40	24.33	4.91	37.64
450 x 450mm	m2	0.55	7.70	23.04	4.61	35.35
600 x 600mm	m2	0.50	7.00	22.78	4.47	34.25

	Unit	Labour	Hours £	Mat'ls £	O & P £	Total £
York stone paving flags spot bedded in cement mortar, straight both ways, jointing in cement mortar brushed in						
50mm thick						
cut to size	m2	0.80	11.20	88.36	14.93	114.49
random sizes	m2	0.90	12.60	73.87	12.97	99.44
63mm thick						
cut to size	m2	0.85	11.90	97.41	16.40	125.71
random sizes	m2	0.95	13.30	82.49	14.37	110.16
75mm thick						
cut to size	m2	0.90	12.60	109.89	18.37	140.86
random sizes	m2	1.00	14.00	92.44	15.97	122.41
Natural granite setts bedded in cement mortar, size 100 × 100 × 150mm	m2	2.75	38.50	32.55	10.66	81.71
Concrete setts bedded in cement mortar, size						
150 × 150 × 80mm	m2	2.75	38.50	31.22	10.46	80.18
200 × 150 × 80mm	m2	2.50	35.00	31.22	9.93	76.15
Concrete paviors, bedded in sand, with sand brushed in, size						
200 × 100 × 60mm	m2	1.00	14.00	10.55	3.68	28.23
200 × 100 × 80mm	m2	1.00	14.00	11.65	3.85	29.50

	Unit	Labour	Hours £	Mat'ls £	O & P £	Total £
FENCING						
Fencing posts						
Galvanised steel posts, size 40 × 40 × 5mm, driven into ground						
900mm high	nr	0.05	0.70	1.66	0.35	2.71
1200mm high	nr	0.05	0.70	1.98	0.40	3.08
1400mm high	nr	0.10	1.40	2.25	0.55	4.20
1800mm high	nr	0.15	2.10	2.78	0.73	5.61
2400mm high	nr	0.15	2.10	3.14	0.79	6.03
Galvanised steel posts, size 40 × 40 × 5mm, with one strut, set in concrete						
900mm high	nr	0.20	2.80	19.34	3.32	25.46
1200mm high	nr	0.20	2.80	22.56	3.80	29.16
1400mm high	nr	0.25	3.50	25.61	4.37	33.48
1800mm high	nr	0.30	4.20	29.01	4.98	38.19
2400mm high	nr	0.30	4.20	33.96	5.72	43.88
Galvanised steel posts, size 40 × 40 × 5mm, with two struts, driven into ground						
900mm high	nr	0.30	4.20	36.53	6.11	46.84
1200mm high	nr	0.30	4.20	39.83	6.60	50.63
1400mm high	nr	0.35	4.90	42.37	7.09	54.36
1800mm high	nr	0.40	5.60	45.19	7.62	58.41
2400mm high	nr	0.40	5.60	51.64	8.59	65.83

	Unit	Labour	Hours £	Mat'ls £	O & P £	Total £
Precast concrete posts size 100 × 100mm, set in concrete						
900mm high	nr	0.25	3.50	19.34	3.43	26.27
1200mm high	nr	0.25	3.50	22.56	3.91	29.97
1400mm high	nr	0.30	4.20	25.61	4.47	34.28
1800mm high	nr	0.35	4.90	29.01	5.09	39.00
2400mm high	nr	0.35	4.90	33.96	5.83	44.69
Precast concrete posts size 100 × 100mm, with one strut, set in concrete						
900mm high	nr	0.40	5.60	42.54	7.22	55.36
1200mm high	nr	0.40	5.60	46.17	7.77	59.54
1400mm high	nr	0.45	6.30	54.66	9.14	70.10
1800mm high	nr	0.55	7.70	58.37	9.91	75.98
2400mm high	nr	0.55	7.70	76.55	12.64	96.89
Precast concrete posts size 100 × 100mm, with two struts, set in concrete						
900mm high	nr	0.50	7.00	58.64	9.85	75.49
1200mm high	nr	0.50	7.00	63.21	10.53	80.74
1400mm high	nr	0.55	7.70	72.36	12.01	92.07
1800mm high	nr	0.60	8.40	83.67	13.81	105.88
2400mm high	nr	0.60	8.40	88.97	14.61	111.98

	Unit	Labour	Hours £	Mat'ls £	O & P £	Total £
Chainlink fencing						
Chainlink fencing, galvanised steel mesh on three line, wires fixed to posts (not included)						
900mm high	m	0.15	2.10	1.55	0.55	4.20
1200mm high	m	0.15	2.10	1.65	0.56	4.31
1400mm high	m	0.20	2.80	2.88	0.85	6.53
1800mm high	m	0.25	3.50	3.54	1.06	8.10
2400mm high	m	0.25	3.50	4.67	1.23	9.40
Chainlink fencing, plastic coated steel mesh on three line, wires fixed to posts (not included)						
900mm high	m	0.15	2.10	1.65	0.56	4.31
1200mm high	m	0.15	2.10	1.75	0.58	4.43
1400mm high	m	0.20	2.80	2.98	0.87	6.65
1800mm high	m	0.25	3.50	3.64	1.07	8.21
2400mm high	m	0.25	3.50	4.77	1.24	9.51
Chestnut fencing						
Chestnut fencing with pales at 75mm centres, fixed to 75 mm posts at 3 metre centres driven into ground						
two line wire						
900mm high	m	0.25	3.50	4.10	1.14	8.74
1100mm high	m	0.30	4.20	4.25	1.27	9.72
1250mm high	m	0.35	4.90	4.56	1.42	10.88

	Unit	Labour	Hours £	Mat'ls £	O & P £	Total £
Chestnut fencing (cont'd)						
three line wire						
900mm high	m	0.25	3.50	4.30	1.17	8.97
1100mm high	m	0.30	4.20	4.45	1.30	9.95
1250mm high	m	0.35	4.90	4.76	1.45	11.11
Strained wire fencing						
Strained wire fencing fixed to concrete posts (not included)						
1000mm high, 3 wires	m	0.10	1.40	0.44	0.28	2.12
1200mm high, 3 wires	m	0.11	1.54	0.44	0.30	2.28
1500mm high, 4 wires	m	0.12	1.68	0.66	0.35	2.69
Barbed wire fencing						
Barbed wire fencing fixed to concrete posts (not included)						
1000mm high, 3 wires	m	0.14	1.96	0.54	0.38	2.88
1200mm high, 3 wires	m	0.16	2.24	0.66	0.44	3.34
1500mm high, 4 wires	m	0.18	2.52	0.78	0.50	3.80

	Unit	Labour	Hours £	Mat'ls £	O & P £	Total £

DRAINAGE

Excavate drain trench by hand, support sides, level and ram bottom of trench, backfill and consolidate with excavated material and and remove surplus to tip for pipes 100mm and 150mm diameter, average depth of trench

	Unit	Labour	Hours £	Mat'ls £	O & P £	Total £
0.50m	m	1.20	16.80	0.00	2.52	19.32
0.75m	m	1.90	26.60	0.00	3.99	30.59
1.00m	m	2.50	35.00	0.00	5.25	40.25
1.25m	m	4.00	56.00	0.00	8.40	64.40
1.50m	m	5.00	70.00	0.00	10.50	80.50
1.75m	m	6.00	84.00	0.00	12.60	96.60
2.00m	m	7.20	100.80	0.00	15.12	115.92
2.25m	m	8.20	114.80	0.00	17.22	132.02
2.50m	m	9.50	133.00	0.00	19.95	152.95
2.75m	m	10.50	147.00	0.00	22.05	169.05
3.00m	m	12.00	168.00	0.00	25.20	193.20

Excavate drain trench by hand, support sides, level and ram bottom of trench, backfill and consolidate with excavated material and and remove surplus to tip for pipes 225mm diameter, average depth of trench

	Unit	Labour	Hours £	Mat'ls £	O & P £	Total £
0.50m	m	1.30	18.20	0.00	2.73	20.93
0.75m	m	2.10	29.40	0.00	4.41	33.81
1.00m	m	2.70	37.80	0.00	5.67	43.47

	Unit	Labour	Hours £	Mat'ls £	O & P £	Total £
Excavate drain trench (cont'd)						
1.25m	m	4.30	60.20	0.00	9.03	69.23
1.50m	m	5.40	75.60	0.00	11.34	86.94
1.75m	m	6.60	92.40	0.00	13.86	106.26
2.00m	m	7.10	99.40	0.00	14.91	114.31
2.25m	m	9.00	126.00	0.00	18.90	144.90
2.50m	m	10.20	142.80	0.00	21.42	164.22
2.75m	m	11.30	158.20	0.00	23.73	181.93
3.00m	m	12.80	179.20	0.00	26.88	206.08
Extra for breaking up by hand						
plain concrete 100mm thi	m2	0.90	12.60	0.00	1.89	14.49
reinforced concrete 100mm thick	m2	1.00	14.00	0.00	2.10	16.10
tarmacadam 75mm thick	m2	0.55	7.70	0.00	1.16	8.86
hardcore 100mm thick	m2	0.40	5.60	0.00	0.84	6.44
soft rock	m3	8.00	112.00	0.00	16.80	128.80
hard rock	m3	10.00	140.00	0.00	21.00	161.00
Sand bed in trench under pipe 100mm diameter, thickness						
100mm	m	0.10	1.40	1.20	0.39	2.99
150mm	m	0.12	1.68	1.80	0.52	4.00
Sand bed in trench under pipe 150mm diameter, thickness						
100mm	m	0.11	1.54	1.80	0.50	3.84
150mm	m	0.13	1.82	2.70	0.68	5.20

	Unit	Labour	Hours £	Mat'ls £	O & P £	Total £
Sand bed in trench under pipe 225mm diameter, thickness						
100mm	m	0.14	1.96	1.90	0.58	4.44
150mm	m	0.16	2.24	2.85	0.76	5.85
Granular filling in bed in trench under pipe 100mm diameter, thickness						
100mm	m	0.12	1.68	2.10	0.57	4.35
150mm	m	0.14	1.96	3.15	0.77	5.88
Granular filling in bed in trench under pipe 150mm diameter, thickness						
100mm	m	0.13	1.82	2.18	0.60	4.60
150mm	m	0.15	2.10	3.27	0.81	6.18
Granular filling in bed in trench under pipe 225mm diameter, thickness						
100mm	m	0.16	2.24	2.33	0.69	5.26
150mm	m	0.18	2.52	3.41	0.89	6.82
Concrete filling in bed in trench under pipe 100mm diameter, thickness						
100mm	m	0.24	3.36	4.05	1.11	8.52
150mm	m	0.28	3.92	6.07	1.50	11.49

	Unit	Labour	Hours £	Mat'ls £	O & P £	Total £
Concrete filling in bed in trench under pipe 150mm diameter, thickness						
100mm	m	0.26	3.64	4.77	1.26	9.67
150mm	m	0.30	4.20	7.15	1.70	13.05
Concrete filling in bed in trench under pipe 225mm diameter, thickness						
100mm	m	0.32	4.48	5.40	1.48	11.36
150mm	m	0.36	5.04	8.10	1.97	15.11
Concrete in bed and haunching to pipe, 100mm diameter, bed thickness						
100mm	m	0.48	6.72	5.40	1.82	13.94
150mm	m	0.56	7.84	8.10	2.39	18.33
Concrete in bed and haunching to pipe, 150mm diameter, bed thickness						
100mm	m	0.52	7.28	6.93	2.13	16.34
150mm	m	0.60	8.40	10.40	2.82	21.62
Concrete in bed and haunching to pipe, 225mm diameter, bed thickness						
100mm	m	0.60	8.40	9.27	2.65	20.32
150mm	m	0.66	9.24	13.90	3.47	26.61

	Unit	Labour	Hours £	Mat'ls £	O & P £	Total £
Granular filling in bed and surround to pipe, 100mm diameter, thickness						
100mm	m	0.36	5.04	2.84	1.18	9.06
150mm	m	0.42	5.88	4.26	1.52	11.66
Granular filling in bed and surround to pipe, 150mm diameter, thickness						
100mm	m	0.40	5.60	3.65	1.39	10.64
150mm	m	0.46	6.44	5.48	1.79	13.71
Granular filling in bed and surround to pipe, 225mm diameter, thickness						
100mm	m	0.44	6.16	3.78	1.49	11.43
150mm	m	0.50	7.00	5.67	1.90	14.57
Concrete filling in bed and surround to pipe, 100mm diameter, thickness						
100mm	m	0.72	10.08	14.22	3.65	27.95
150mm	m	0.84	11.76	21.33	4.96	38.05
Concrete filling in bed and surround to pipe, 150mm diameter, thickness						
100mm	m	0.72	10.08	18.27	4.25	32.60
150mm	m	0.84	11.76	27.41	5.88	45.05

	Unit	Labour	Hours £	Mat'ls £	O & P £	Total £
Concrete filling in bed and surround to pipe, 225mm diameter, thickness						
100mm	m	0.72	10.08	18.90	4.35	33.33
150mm	m	0.84	11.76	29.90	6.25	47.91
Hepworths' Supersleve vitrified clay drain pipes, spigot and socket joints with sealing rings, 100mm diameter						
laid in trenches	m	0.36	5.04	6.39	1.71	13.14
in lengths not exceeding 3m	m	0.54	7.56	6.39	2.09	16.04
bend	nr	0.30	4.20	8.70	1.94	14.84
rest bend	nr	0.30	4.20	13.53	2.66	20.39
single junction	nr	0.30	4.20	12.76	2.54	19.50
Hepworths' Supersleve vitrified clay drain pipes, spigot and socket joints with sealing rings, 150mm diameter						
laid in trenches	m	0.60	8.40	12.19	3.09	23.68
in lengths not exceeding 3m	m	0.40	5.60	12.19	2.67	20.46
bend	nr	0.30	4.20	12.16	2.45	18.81
rest bend	nr	0.30	4.20	15.63	2.97	22.80
single junction	nr	0.30	4.20	16.29	3.07	23.56

	Unit	Labour	Hours £	Mat'ls £	O & P £	Total £
Vitrified clay inlet gulley complete with grid and surrounded with concrete	nr	1.50	21.00	40.58	9.24	70.82
Vitrified clay back inlet gulley complete with grid and surrounded with concrete	nr	1.50	21.00	57.99	11.85	90.84
Vitrified clay paved area gulley complete with grid and surrounded with concrete	nr	1.50	21.00	47.17	10.23	78.40
Manholes						
Excavate by hand for manhole not exceeding						
1.0m deep	m3	4.00	56.00	0.00	8.40	64.40
1.5m deep	m3	4.50	63.00	0.00	9.45	72.45
2.0m deep	m3	5.60	78.40	0.00	11.76	90.16
Load surplus excavated material into barrows wheel 50m and deposit into skip	m3	2.20	30.80	0.00	4.62	35.42
Earthwork support not exceeding 2m between opposing faces, depth not exceeding						
1.0m deep	m2	0.35	4.90	1.92	1.02	7.84
2.0m deep	m2	0.40	5.60	2.22	1.17	8.99
4.0m deep	m2	0.45	6.30	2.51	1.32	10.13

	Unit	Labour	Hours £	Mat'ls £	O & P £	Total £
Site-mixed concrete in base of manhole, thickness						
100-150mm	m3	2.00	28.00	94.88	18.43	141.31
150-300mm	m3	1.80	25.20	94.88	18.01	138.09
Site-mixed concrete in benching to manhole, average thickness 225mm	m3	6.00	84.00	94.88	26.83	205.71
Engineering bricks Class 'B' in cement mortar in walls of manhole	m2	4.80	67.20	57.86	18.76	143.82
Extra for fair face and flush pointing	m2	0.25	3.50	0.00	0.53	4.03
Build in ends of pipes to one brick wall and make good, pipe diameter						
100mm	nr	0.15	2.10	0.00	0.32	2.42
150mm	nr	0.19	2.65	0.00	0.40	3.04
Galvanised step iron built into brickwork	nr	0.20	2.80	5.42	1.23	9.45
Cast iron manhole cover, frame, bedded in cement mortar						
Grade A, light duty, size 600 × 450mm	nr	1.90	26.60	71.64	14.74	112.98
Grade B, medium duty, size 600 × 450mm	nr	2.00	28.00	114.69	21.40	164.09

	Unit	Labour	Hours £	Mat'ls £	O & P £	Total £
Best quality vitrified clay channels, bedded in cement mortar						
half-section straight main channel						
100mm diameter × 300mm	nr	0.75	10.50	2.94	2.02	15.46
100mm diameter × 600mm	nr	0.85	11.90	5.33	2.58	19.81
100mm diameter × 1000mm	nr	0.95	13.30	6.84	3.02	23.16
150mm diameter × 300mm	nr	0.75	10.50	4.55	2.26	17.31
150mm diameter × 600mm	nr	0.85	11.90	6.79	2.80	21.49
150mm diameter × 1000mm	nr	0.95	13.30	9.44	3.41	26.15
Half-section 90 degrees channel bend						
100mm	nr	0.30	4.20	5.67	1.48	11.35
150mm	nr	0.40	5.60	8.61	2.13	16.34
Three-quarter section 90° channel bend						
100mm	nr	1.20	16.80	12.04	4.33	33.17
150mm	nr	1.30	18.20	21.49	5.95	45.64

	Unit	Labour	Hours £	Mat'ls £	O & P £	Total £
Land drainage						
Excavate trench 225mm wide by hand, remove excavated material, average depth						
500mm	m	0.35	4.90	0.00	0.74	5.64
750mm	m	0.50	7.00	0.00	1.05	8.05
1000mm	m	0.75	10.50	0.00	1.58	12.08
Excavate trench 300mm wide by hand, remove excavated material, average depth						
500mm	m	0.45	6.30	0.00	0.95	7.25
750mm	m	0.70	9.80	0.00	1.47	11.27
1000mm	m	0.90	12.60	0.00	1.89	14.49
Agricultural clay field drain pipe, 300mm long, laid butt jointed in trench, diameter						
75mm	m	0.11	1.54	2.68	0.63	4.85
100mm	m	0.12	1.68	4.57	0.94	7.19
150mm	m	0.14	1.96	9.04	1.65	12.65
Single junction						
75mm	nr	0.06	0.84	9.98	1.62	12.44
100mm	nr	0.07	0.98	12.61	2.04	15.63
150mm	nr	0.08	1.12	15.56	2.50	19.18

	Unit	Labour	Hours £	Mat'ls £	O & P £	Total £
Flexible plastic perforated field drain pipe, diameter						
60mm	m	0.02	0.28	0.28	0.08	0.64
80mm	m	0.02	0.28	0.42	0.11	0.81
100mm	m	0.03	0.42	0.68	0.17	1.27
Single junction						
60 × 60mm	nr	0.06	0.84	1.47	0.35	2.66
60 × 80mm	nr	0.06	0.84	1.59	0.36	2.79
60 × 100mm	nr	0.06	0.84	1.63	0.37	2.84
80 × 80mm	nr	0.06	0.84	1.58	0.36	2.78
80 × 100mm	nr	0.06	0.84	1.70	0.38	2.92
Backfilling field trench with gravel, blinded with sand and replace with turf laid aside						
trench width 225mm						
depth 500mm	m	0.03	0.42	2.02	0.37	2.81
depth 750mm	m	0.03	0.42	3.00	0.51	3.93
depth 1000mm	m	0.03	0.42	3.98	0.66	5.06
trench width 300mm						
depth 500mm	m	0.03	0.42	2.70	0.47	3.59
depth 750mm	m	0.03	0.42	3.95	0.66	5.03
depth 1000mm	m	0.03	0.42	5.21	0.84	6.47

Part Three

APPROXIMATE ESTIMATING

Brick walling

Masonry

Kerbs and edgings

Beds and pavings

Drainage

	Unit	Rate £

BRICK WALLING

Excavate trench, lay concrete foundation, build wall 1.5m high in common bricks (basic price £140 per 1000) in cement mortar (1:3), stretcher bond

	Unit	Rate £
half brick thick	m	85.00
half brick thick, curved	m	100.00
one brick thick	m	130.00
one thick, curved	m	145.00
one and a half brick thick	m	170.00
two brick thick	m	205.00

Excavate trench, lay concrete foundation, build wall 1.5m high in common bricks (basic price £140 per 1000) in cement mortar (1:3), stretcher bond, facework one side

	Unit	Rate £
half brick thick	m	88.00
half brick thick, curved	m	103.00
one brick thick	m	133.00
one thick, curved	m	148.00
one and a half brick thick	m	173.00
two brick thick	m	208.00

Excavate trench, lay concrete foundation, build wall 1.5m high in common bricks (basic price £140 per 1000) in cement mortar (1:3), stretcher bond, facework both sides

	Unit	Rate £
half brick thick	m	90.00
half brick thick, curved	m	105.00

	Unit	Rate £
Brick walling (cont'd)		
one brick thick	m	135.00
one thick, curved	m	150.00
one and a half brick thick	m	175.00
two brick thick	m	210.00
Excavate trench, lay concrete foundation, build wall 1.5m high in common bricks (basic price £200 per 1000) in cement mortar (1:3), stretcher bond		
half brick thick	m	95.00
half brick thick, curved	m	110.00
one brick thick	m	150.00
one thick, curved	m	165.00
one and a half brick thick	m	195.00
two brick thick	m	210.00
Excavate trench, lay concrete foundation, build wall 1.5m high in common bricks (basic price £200 per 1000) in cement mortar (1:3), stretcher bond, facework one side		
half brick thick	m	98.00
half brick thick, curved	m	113.00
one brick thick	m	153.00
one thick, curved	m	168.00
one and a half brick thick	m	198.00
two brick thick	m	213.00

	Unit	Rate £
Excavate trench, lay concrete foundation, build wall 1.5m high in common bricks (basic price £200 per 1000) in cement mortar (1:3), stretcher bond, facework both sides		
half brick thick	m	100.00
half brick thick, curved	m	115.00
one brick thick	m	155.00
one thick, curved	m	170.00
one and a half brick thick	m	200.00
two brick thick	m	215.00
Excavate trench, lay concrete foundation, build wall 1.5m high in common bricks (basic price £250 per 1000) in cement mortar (1:3), stretcher bond		
half brick thick	m	113.00
half brick thick, curved	m	128.00
one brick thick	m	208.00
one thick, curved	m	218.00
one and a half brick thick	m	248.00
two brick thick	m	308.00
Excavate trench, lay concrete foundation, build wall 1.5m high in common bricks (basic price £250 per 1000) in cement mortar (1:3), stretcher bond, facework one side		
half brick thick	m	115.00
half brick thick, curved	m	130.00
one brick thick	m	210.00
one thick, curved	m	220.00

	Unit	Rate £
Brick walling (cont'd)		
one and a half brick thick	m	250.00
two brick thick	m	310.00
Excavate trench, lay concrete foundation, build wall 1.5m high in common bricks (basic price £350 per 1000) in cement mortar (1:3), stretcher bond, facework one side		
half brick thick	m	120.00
half brick thick, curved	m	135.00
one brick thick	m	195.00
one thick, curved	m	210.00
one and a half brick thick	m	250.00
two brick thick	m	325.00
Excavate trench, lay concrete foundation, build wall 1.5m high in facing bricks (basic price £350 per 1000) in cement mortar (1:3), English garden wall bond, facework both sides		
half brick thick	m	125.00
half brick thick, curved	m	140.00
one brick thick	m	200.00
one thick, curved	m	215.00
one and a half brick thick	m	255.00
two brick thick	m	245.00
Excavate trench, lay concrete foundation, build wall 1.5m high in facing bricks (basic price £500 per 1000) in cement mortar (1:3), English garden wall bond, facework one side		

	Unit	Rate £
half brick thick	m	130.00
half brick thick, curved	m	145.00
one brick thick	m	215.00
one thick, curved	m	230.00
one and a half brick thick	m	290.00
two brick thick	m	325.00
Excavate trench, lay concrete foundation, build wall 1.5m high in facing bricks (basic price £500 per 1000) in cement mortar (1:3), English garden wall bond, facework both sides		
half brick thick	m	135.00
half brick thick, curved	m	150.00
one brick thick	m	220.00
one thick, curved	m	235.00
one and a half brick thick	m	295.00
two brick thick	m	330.00
Excavate trench, lay concrete foundation, build wall 1.5m high in facing bricks (basic price £250 per 1000) in cement mortar (1:3), Flemish bond, facework one side		
half brick thick	m	110.00
half brick thick, curved	m	125.00
one brick thick	m	205.00
one thick, curved	m	220.00
one and a half brick thick	m	245.00
two brick thick	m	305.00

	Unit	Rate £
Brick walling (cont'd)		
Excavate trench, lay concrete foundation, build wall 1.5m high in facing bricks (basic price £250 per 1000) in cement mortar (1:3), Flemish bond, facework both sides		
half brick thick	m	115.00
half brick thick, curved	m	130.00
one brick thick	m	210.00
one thick, curved	m	225.00
one and a half brick thick	m	250.00
two brick thick	m	310.00
Excavate trench, lay concrete foundation, build wall 1.5m high in facing bricks (basic price £350 per 1000) in cement mortar (1:3), Flemish bond, facework one side		
half brick thick	m	120.00
half brick thick, curved	m	135.00
one brick thick	m	195.00
one thick, curved	m	210.00
one and a half brick thick	m	260.00
two brick thick	m	325.00
Excavate trench, lay concrete foundation, build wall 1.5m high in facing bricks (basic price £350 per 1000) in cement mortar (1:3), Flemish bond, facework both sides		
half brick thick	m	125.00
half brick thick, curved	m	140.00

	Unit	Rate £
one brick thick	m	200.00
one thick, curved	m	215.00
one and a half brick thick	m	265.00
two brick thick	m	330.00

Excavate trench, lay concrete foundation, build wall 1.5m high in facing bricks (basic price £500 per 1000) in cement mortar (1:3), Flemish bond, facework one side

	Unit	Rate £
half brick thick	m	130.00
half brick thick, curved	m	155.00
one brick thick	m	215.00
one thick, curved	m	230.00
one and a half brick thick	m	290.00
two brick thick	m	370.00

Excavate trench, lay concrete foundation, build wall 1.5m high in facing bricks (basic price £500 per 1000) in cement mortar (1:3), Flemish bond, facework both sides

	Unit	Rate £
half brick thick	m	135.00
half brick thick, curved	m	160.00
one brick thick	m	220.00
one thick, curved	m	235.00
one and a half brick thick	m	295.00
two brick thick	m	375.00

	Unit	Rate £
MASONRY		
Excavate trench, lay concrete foundation, build random rubble wall laid dry, 1.5m high, thickness		
300mm	m	185.00
450mm	m	215.00
500mm	m	240.00
Excavate trench, lay concrete foundation, build random rubble wall battered one side, laid dry, 1.5m high, thickness		
300mm	m	190.00
450mm	m	245.00
500mm	m	270.00
Excavate trench, lay concrete foundation, build random rubble wall battered both sides, laid dry, 1.5m high, thickness		
300mm	m	195.00
450mm	m	250.00
500mm	m	275.00
Excavate trench, lay concrete foundation, build random rubble wall in gauged mortar (1:1:6), 1.5m high, thickness		
300mm	m	195.00
450mm	m	250.00
500mm	m	275.00

	Unit	Rate £
Excavate trench, lay concrete foundation, build random rubble wall in gauged mortar (1:1:6), 1.5m high, battered one side, thickness		
300mm	m	200.00
450mm	m	260.00
500mm	m	280.00
Excavate trench, lay concrete foundation, build random rubble wall in gauged mortar (1:1:6), 1.5m high, battered both sides, thickness		
300mm	m	215.00
450mm	m	275.00
500mm	m	295.00
Excavate trench, lay concrete foundation, build irregular coursed rubble wall in gauged mortar (1:1:6), 1.5m high, thickness		
300mm	m	170.00
450mm	m	235.00
500mm	m	265.00
Excavate trench, lay concrete foundation, build coursed rubble wall in gauged mortar (1:1:6), 1.5m high, thickness		
300mm	m	175.00
450mm	m	240.00
500mm	m	270.00

	Unit	Rate £
KERBS AND EDGINGS		
Excavate trench, lay concrete foundation for precast concrete straight kerb, size		
127 × 254mm	m	20.00
152 × 305mm	m	20.00
Excavate trench, lay concrete foundation for precast concrete curved kerb, size		
127 × 254mm	m	22.00
152 × 305mm	m	28.00
Excavate trench, lay concrete foundation for precast concrete straight channel, size		
127 × 254mm	m	22.00
Excavate trench, lay concrete foundation for precast concrete curved channel, size		
127 × 254mm	m	23.00
Excavate trench, lay concrete foundation for precast concrete straight edging, size		
51 × 152mm	m	20.00
52 × 203mm	m	20.00

	Unit	Rate £
BEDS AND PAVINGS		
Excavate by hand, lay sub-base of		
granular material 100mm thick, reinforced concrete bed 100mm thick	m2	32.00
granular material 150mm thick, reinforced concrete bed 100mm thick	m2	34.00
sand 100mm thick, reinforced concrete bed 100mm thick	m2	32.00
sand 150mm thick, reinforced concrete bed 150mm thick	m2	34.00
hardcore 100mm thick, reinforced concrete bed 100mm thick	m2	32.00
hardcore 150mm thick, reinforced concrete bed 100mm thick	m2	34.00
Excavate by hand, lay sub-base of sand 100mm thick, lay brick paving 215 × 103mm (£250 per 1000)		
straight joints both ways		
bricks laid flat	m2	56.00
bricks laid on edge	m2	75.00
herringbone pattern		
bricks laid flat	m2	60.00
bricks laid on edge	m2	80.00
Excavate by hand, lay sub-base of sand 100mm thick, lay brick paving 215 × 103mm (£350 per 1000)		

	Unit	Rate £
straight joints both ways		
bricks laid flat	m2	62.00
bricks laid on edge	m2	80.00
herringbone pattern		
bricks laid flat	m2	66.00
bricks laid on edge	m2	85.00
Excavate by hand, lay sub-base of sand 100mm thick, lay brick paving 215 × 103mm (£500 per 1000)		
straight joints both ways		
bricks laid flat	m2	70.00
bricks laid on edge	m2	88.00
herringbone pattern		
bricks laid flat	m2	75.00
bricks laid on edge	m2	92.00
Excavate by hand, lay sub-base of sand 100mm thick, lay precast precast concrete flags, spot bedded cement mortar		
natural colour		
450 × 450 × 50mm	m2	40.00
450 × 450 × 50mm	m2	36.00
600 × 600 × 50mm	m2	32.00
600 × 900 × 50mm	m2	30.00
coloured		
450 × 450 × 50mm	m2	44.00
450 × 450 × 50mm	m2	40.00
600 × 600 × 50mm	m2	38.00
600 × 900 × 50mm	m2	36.00

	Unit	Rate £
Excavate by hand, lay sub-base of sand 100mm thick, lay reconstituted York stone flags, spot bedded cement mortar		
300 × 300 × 50mm	m2	60.00
300 × 450 × 50mm	m2	55.00
300 × 600 × 50mm	m2	50.00
450 × 450 × 50mm	m2	48.00
600 × 600 × 50mm	m2	46.00
Excavate by hand, lay sub-base of sand 100mm thick, lay York stone flags, spot bedded in cement mortar, laid straight both ways		
50mm thick		
cut to size	m2	130.00
random sizes	m2	115.00
63mm thick		
cut to size	m2	140.00
random sizes	m2	125.00
75mm thick		
cut to size	m2	155.00
random sizes	m2	135.00
Excavate by hand, lay sub-base of sand 100mm thick, lay natural granite setts, size 100 × 100 × 150mm, bedded in cement mortar	m2	95.00

	Unit	Rate £
Excavate by hand, lay sub-base of sand 100mm thick, lay concrete setts, bedded in cement mortar, size		
150 × 150 × 80mm	m2	95.00
200 × 150 × 80mm	m2	90.00
Excavate by hand, lay sub-base of sand 100mm thick, lay concrete paviors, bedded in cement mortar, size		
200 × 100 × 60mm	m2	40.00
200 × 100 × 80mm	m2	42.00

DRAINAGE

	Unit	Rate £
Excavate trench by hand, lay 100mm diameter 'Supersleve' pipe, granular bed and surround, trench depth		
0.50m	m	47.00
0.75m	m	57.00
1.00m	m	67.00
1.25m	m	81.00
1.50m	m	107.00
1.75m	m	123.00
2.00m	m	142.00
2.25m	m	159.00
2.50m	m	179.00
2.75m	m	196.00
3.00m	m	220.00

	Unit	Rate £
Excavate trench by hand, lay 150mm diameter 'Hepsleve' pipe, granular bed and surround, trench depth		
0.50m	m	57.00
0.75m	m	67.00
1.00m	m	77.00
1.25m	m	91.00
1.50m	m	117.00
1.75m	m	133.00
2.00m	m	152.00
2.25m	m	169.00
2.50m	m	189.00
2.75m	m	206.00
3.00m	m	230.00
Manhole complete including hand excavation, concrete base and benching, engineering brickwork, channels and cast iron cover, depth		
1.00m	nr	495.00
1.50m	nr	640.00
2.00m	nr	780.00

Land drainage

	Unit	Rate £
Excavate trench 225mm wide for 75mm clay field drain pipe and backfill with gravel		
trench depth, 500mm	m	15.00
trench depth, 750mm	m	18.00
trench depth, 1000mm	m	22.00

	Unit	Rate £
Excavate trench 300mm wide for 75mm clay field drain pipe and backfill with gravel		
trench depth, 500mm	m	17.00
trench depth, 750mm	m	21.00
trench depth, 1000mm	m	25.00
Excavate trench 225mm wide for 100mm clay field drain pipe and backfill with gravel		
trench depth, 500mm	m	18.00
trench depth, 750mm	m	21.00
trench depth, 1000mm	m	25.00
Excavate trench 300mm wide for 100mm clay field drain pipe and backfill with gravel		
trench depth, 500mm	m	20.00
trench depth, 750mm	m	24.00
trench depth, 1000mm	m	28.00
Excavate trench 225mm wide for 100mm clay field drain pipe and backfill with gravel		
trench depth, 500mm	m	23.00
trench depth, 750mm	m	26.00
trench depth, 1000mm	m	30.00

	Unit	Rate £
Excavate trench 300mm wide for 100mm clay field drain pipe and backfill with gravel		
trench depth, 500mm	m	25.00
trench depth, 750mm	m	29.00
trench depth, 1000mm	m	33.00
Excavate trench 225mm wide for 60mm perforated field drain pipe and backfill with gravel		
trench depth, 500mm	m	11.00
trench depth, 750mm	m	14.00
trench depth, 1000mm	m	18.00
Excavate trench 300mm wide for 80mm perforated field drain pipe and backfill with gravel		
trench depth, 500mm	m	13.00
trench depth, 750mm	m	17.00
trench depth, 1000mm	m	21.00
Excavate trench 225mm wide for 100mm clay field drain pipe and backfill with gravel		
trench depth, 500mm	m	14.00
trench depth, 750mm	m	17.00
trench depth, 1000mm	m	21.00
Excavate trench 300mm wide for 100mm clay field drain pipe and backfill with gravel		

	Unit	Rate £
Land drainage (cont'd)		
trench depth, 500mm	m	16.00
trench depth, 750mm	m	20.00
trench depth, 1000mm	m	24.00
Excavate trench 225mm wide for 100mm clay field drain pipe and backfill with gravel		
trench depth, 500mm	m	19.00
trench depth, 750mm	m	22.00
trench depth, 1000mm	m	26.00
Excavate trench 300mm wide for 100mm clay field drain pipe and backfill with gravel		
trench depth, 500mm	m	21.00
trench depth, 750mm	m	25.00
trench depth, 1000mm	m	29.00

Part Four

TOOL AND EQUIPMENT HIRE

	24 hours	Extra 24 hours	Week
	£	£	£

TOOL AND EQUIPMENT HIRE

The following rates are based on average hire charges made by hire firms in the UK. Check your local dealer for more information. The rates exclude VAT.

Alloy towers

	24 hours £	Extra 24 hours £	Week £
Single width, height			
2.2m	41.00	22.00	82.00
3.2m	51.00	26.00	102.00
4.2m	60.00	30.00	120.00
5.2m	70.00	35.00	140.00
6.2m	78.00	39.00	156.00
7.2m	87.00	43.00	174.00
8.2m	96.00	48.00	192.00
9.2m	104.00	52.00	208.00
10.2m	112.00	56.00	224.00
Full width, height			
2.2m	76.00	38.00	152.00
3.2m	88.00	44.00	176.00
4.2m	100.00	50.00	200.00
5.2m	110.00	55.00	220.00
6.2m	124.00	62.00	248.00
7.2m	148.00	74.00	296.00
8.2m	160.00	80.00	320.00

	24 hours £	Extra 24 hours £	Week £
Ladders			
Roof ladders			
single, 5.9m	24.00	12.00	48.00
double, 4.6m	22.00	11.00	44.00
double, 7.6m	26.00	13.00	46.00
Push-up ladders			
double, 3.5m	16.00	8.00	32.00
double, 5.0m	22.00	11.00	44.00
treble, 2.5m	16.00	8.00	32.00
treble, 3.5m	22.00	11.00	44.00
Rope-operated ladders			
double, 6.0m	38.00	19.00	76.00
treble, 5.2m	44.00	22.00	88.00
treble, 6.0m	50.00	25.00	100.00
Stand off ladder stay	6.00	3.00	12.00
Compaction			
Vibrating rollers			
large	80.00	40.00	155.00
small	60.00	30.00	120.00
Rammer	45.00	25.00	90.00
Plate compactor			
diesel	70.00	35.00	140.00
petrol	65.00	35.00	135.00

	24 hours	Extra 24 hours	Week
	£	£	£
Concrete work			
Tip up			
electric	12.00	6.00	25.00
petrol	15.00	10.00	30.00
Beam screed	70.00	35.00	130.00
Power trowel	45.00	25.00	90.00
Poker vibrator			
electric	30.00	15.00	60.00
petrol	50.00	25.00	100.00
Pumping			
Diaphragm pumps			
50mm	55.00	30.00	110.00
75mm	65.00	35.00	130.00
Centrifugal pumps			
50mm	45.00	25.00	90.00
75mm	50.00	30.00	100.00
Drainage			
Drain snake			
powered	55.00	30.00	110.00
man held	25.00	12.00	50.00
Power jet	175.00	90.00	350.00
Drain inspection camera	80.00	45.00	160.00

	24 hours	Extra 24 hours	Week
	£	£	£
Landscaping			
Lawn aerator			
powered	70.00	35.00	140.00
man held	15.00	8.00	30.00
Roller	10.00	5.00	20.00
Uniseeder	100.00	50.00	200.00
Mower			
rotary (840mm)	100.00	50.00	200.00
rotary with box (525mm)	30.00	15.00	60.00
Scarifier	40.00	20.00	80.00
Flailer	80.00	40.00	160.00
Flame gun	20.00	10.00	40.00
Brush cutter (ride on)	140.00	70.00	280.00
Turf cutter	90.00	45.00	180.00
Power digger	45.00	20.00	80.00
Tiller	20.00	10.00	40.00
Stump grinder	110.00	55.00	220.00
Shredder			
45mm	35.00	20.00	70.00
150mm	180.00	90.00	360.00

	24 hours	Extra 24 hours	Week
	£	£	£
Hedge trimmer			
electric	12.00	6.00	24.00
petrol	35.00	18.00	70.00
Long handled trimmer			
1.3m	40.00	20.00	80.00
2.4m	45.00	22.00	90.00
Logging saw	40.00	20.00	80.00
Pruning saw	20.00	10.00	40.00
Long handle pruner			
petrol	40.00	20.00	80.00
man held	10.00	5.00	20.00
Chain saw			
petrol	60.00	30.00	120.00
electric	50.00	24.00	100.00
Leaf sucker			
petrol	30.00	15.00	60.00
electric	10.00	5.00	20.00
Fence post borer			
petrol	50.00	25.00	100.00
man held	10.00	5.00	20.00
Sundries			
Tarpaulins			
4 × 5m			22.00
8 × 5m			28.00

Part Five

BUSINESS MATTERS

Starting a business

Running a business

Taxation

Starting a business

Most small businesses come into being for one of two reasons – ambition or desperation! A person with genuine ambition for commercial success will never be completely satisfied until he has become self-employed and started his own business. But many successful businesses have been started because the proprietor was forced into this course of action because of redundancy.

Before giving up his job, the would-be businessman should consider carefully whether he has the required skills and the temperament to survive in the highly competitive self-employed market. Before commencing in business it is essential to assess the commercial viability of the intended business because it is pointless to finance a business that is not going to be commercially viable.

In the early stages it is important to make decisions such as: What exactly is the product being sold? What is the market view of that product? What steps are required before the developed product is first sold and where are those sales coming from?

As much information as possible should be obtained on how to run a business before taking the plunge. Sales targets should be set and it should be clearly established how those important first sales are obtained. Above all, do not underestimate the amount of time required to establish and finance a new business venture.

Whatever the size of the business it is important that you put in writing exactly what you are trying to do. This means preparing a business plan that will not only assist in establishing your business aims but is essential if you need to raise finance. The contents of a typical business plan are set out later. It is important to realise that you are not on your own and there are many contacts and advising agencies that can be of assistance.

Potential customers and trade contacts

Many persons intending to start a business in the construction industry will have already had experience as employees. Use all contacts to check the market, establish the sort of work that is available and the current charge-out rates.

In the domestic market, check on the competition for prices and services provided. Study advertisements for your kind of work and try to get firm promises of work before the start-up date.

Testing the market

Talk to as many traders as possible operating in the same field. Identify if the market is in the industrial, commercial, local government or in the domestic field. Talk to prospective customers and clients and consider how you can improve on what is being offered in terms of price, quality, speed, convenience, reliability and back-up service.

Business links

There is no shortage of information about the many aspects of starting and running your own business. Finance, marketing, legal requirements, developing your business idea and taxation matters are all the subject of a mountain of books, pamphlets, guides and courses so it should not be necessary to pay out a lot of money for this information. Indeed, the likelihood is that the aspiring businessman will be overwhelmed with information and will need professional guidance to reduce the risk of wasting time on studying unnecessary subjects.

Business Links are now well established and provide a good place to start for both information and advice. These organisations provide a 'one-stop-shop' for advice and assistance to owner-managed businesses. They will often replace the need to contact Training and Enterprise Councils (TECs) and many of the other official organisations listed below.
Point of contact: telephone directory for address.

Training and Enterprise Councils (TECs)

TECs are comprised of a board of directors drawn from the top men in local industry, commerce, education, trade unions etc., who, together with their staff and experienced business counsellors, assist both new and established concerns in all aspects of running a business. This takes the form of across-the-table advice and also hands-on assistance in management, marketing and finance if required. There are also training courses and seminars available in most areas together with the possibility of grants in some areas.
Point of contact: local Jobcentre or Citizens' Advice Bureau.

Banks

Approach banks for information about the business accounts and financial services that are available. Your local Business Link can advise on how best to find a suitable bank manager and inform you as to what the bank will require.

Shop around several banks and branches if you are not satisfied at first because managers vary widely in their views on what is a viable business proposition. Remember, most banks have useful free information packs to help business start-up.

Point of contact: local bank manager.

HM Inspector of Taxes

Make a preliminary visit to the local tax office enquiry counter for their publications on income tax and national insurance contributions.

SA/Bk 3	Self assessment. A guide to keeping records for the self employed
IR 15(CIS)	Construction Industry Tax Deduction Scheme
CWL	Starting your own business,
IR 40(CIS)	Conditions for Getting a Sub-Contractor's Tax Certificate
NE1	PAYE for Employers (if you employ someone)
NE3	PAYE for new and small Employers
IR 56/N139	Employed or Self-Employed. A guide for tax and National Insurance
CA02	National Insurance contributions for self employed people with small earnings.

Remember, the onus is on the taxpayer, within three months, to notify the Inland Revenue that he is in business and failure to do so may result in the imposition of £100 penalty. Either send a letter or use the form provided at the back of the *'Starting your own business booklet'* to the Inland Revenue National Insurance Contributions Office and they will inform your local tax office of the change in your employment status.

Point of contact: telephone directory for address.

Inland Revenue National Insurance Contributions Office

Self Employment Services
Customer Accounts Section
Longbenton
Newcastle NE 98 1ZZ

Telephone the Call Centre on 0845 9154655 and ask for the following publications:

CWL2	Class 2 and Class 4 Contributions for the Self Employed
CA02	People with Small Earnings from Self-Employment
CA04	Direct Debit - The Easy Way to Pay. Class 2 and Class 3
CA07	Unpaid and Late Paid Contributions and for Employers
CWG1	Employer's Quick Guide to PAYE and NIC Contributions
CA30	Employer's Manual to Statutory Sick Pay

VAT

The VAT office also offer a number of useful publications, including;

700	The VAT Guide
700/1	Should I be Registered for VAT?
731	Cash Accounting
732	Annual Accounting
742	Land and Property

Information about the Cash Accounting Scheme and the introduction of annual VAT returns are dealt with later.
Point of contact: telephone directory for address.

Local authorities

Authorities vary in provisions made for small businesses but all have been asked to simplify and cut delays in planning applications. In Assisted Areas, rent-free periods and reductions in rates may be available on certain industrial and commercial properties. As a preliminary to either purchasing or renting business premises, the following booklets will be helpful:

Step by Step Guide to Planning Permission for Small Businesses, and *Business Leases and Security of Tenure*

Both are issued by the Department of Employment and are available at council offices, Citizens' Advice Bureau and TEC offices. Some authorities run training schemes in conjunction with local industry and educational establishments.

Point of contact: usually the Planning Department - ask for the Industrial Development or Economic Development Officer.

Department of Trade and Industry

The services formally provided by the Department are now increasingly being provided by Business Link . The Department can still, however, provide useful information on available grants for start-ups.
Point of contact: telephone 0207-215 5000 and ask for the address and telephone number of the nearest DTI office and copies of their explanatory booklets.

Department of Transport and the Regions

Regulations are now in force relating to all forms of waste other than normal household rubbish. Any business that produces, stores, treats, processes, transports, recycles or disposes of such waste has a 'duty of care' to ensure it is properly discarded and dealt with.

Practical guidance on how to comply with the law (it is a criminal offence punishable by a fine not to) is contained in a booklet *Waste Management: The Duty of Care: A Code of Practice,* obtainable from HMSO Publication Centre, PO Box 276, London SW8 5DT. Telephone 0207-873 9090.

Accountant

The services of an accountant are to be strongly recommended from the beginning because the legal and taxation requirements start immediately and must be properly complied with if trouble is to be avoided later. A qualified accountant must be used if a limited company is being formed but an accountant will give advice on a whole range of business issues including book-keeping, tax planning and compliance to finance raising and will help in preparing annual accounts.

It is worth spending some time finding an accountant who has other clients in the same line of business and is able to give sound advice particularly on taxation and business finance and is not so overworked that damaging delays in producing accounts are likely to arise. Ask other traders whether they can recommend their own accountant. Visit more than one firm of accountants, ask about the fees they charge and how much the production of annual accounts and agreement with the Inland Revenue are likely to cost. A good accountant is worth every penny of his fees and will save you both money and worry.

Solicitor

Many businesses operate without the services of a solicitor but there are a number of occasions when legal advice should be sought. In particular, no-one should sign a lease of premises without taking legal advice because a business can encounter financial difficulty through unnoticed liabilities in its lease. Either an accountant or solicitor will help with drawing up a partnership agreement that all partnerships should have. A solicitor will also help to explain complex contractual terms and prepare draft contracts if the type of business being entered into requires them.

Insurance broker

Policies are available to cover many aspects of business including:

- employer's liability - compulsory if the business has employees
- public liability - essential in the construction industry
- motor vehicles
- theft of stock, plant and money
- fire and storm damage
- personal accident and loss of profits
- key man cover.

Brokers are independent advisers who will obtain competitive quotations on your behalf. See more than one broker before making a decision - their advice is normally given free and without obligation.
Point of contact: telephone directory or write for a list of local members to:

The British Insurance Brokers' Association
Consumer Relations Department
BIBA House
14 Bevis Marks
London
EC3A 7NT (telephone: 0207-623 9043)

or contact
The Association of British Insurers
51 Gresham Street
London
EC2V 7HQ (telephone: 0207-600 3333)

who will supply free a package of very useful advice files specially designed for the small business.

The Health and Safety Executive

The Executive operates the legislation covering everyone engaged in work activities and has issued a very useful set of '*Construction Health Hazard Information Sheets*' covering such topics as handling cement, lead and solvents, safety in the use of ladders, scaffolding, hoists, cranes, flammable liquids, asbestos, roofs and compressed gases etc. A pack of these may be obtained free from your local HSE office or The Health & Safety Executive Central Office, Sheffield (telephone: (01142-892345) or HSE Publications (telephone: 01787-881165).

Business plan

As stated before, once the relevant information has been obtained it should be consolidated into a formal business plan. The complexity of the plan will depend in the main on the size and nature of the business concerned. Consideration should be given to the following points.

Objectives

It is important to establish what you are trying to achieve both for you and the business. A provider of finance may be particularly influenced by your ability to achieve short- and medium-term goals and may have confidence in continuing to provide finance for the business. From an individual point of view, it is important to establish goals because there is little point in having a business that only serves to achieve the expectations of others whilst not rewarding the would-be businessman.

History

If you already own an existing business then commentary on its existing background structure and history to date can be of assistance. There is no substitute for experience and any existing contacts you have in the construction industry will be of assistance to you. The following points should also be considered for inclusion:

- a brief history of the business identifying useful contacts made
- the development of the business, highlighting significant successes and their relevance to the future
- principal reasons for taking the decision to pursue this new venture
- details of present financing of the business.

Products or services

It is important to establish precisely what it is you are going to sell. Does the product or service have any unique qualities which gives it your advantages over competitors? For example, do you have an ability to react more quickly than your competitors and are you perceived to deliver a higher quality product or service? A typical business plan would include:

- description of the main products and services
- statement of disadvantages and advising how they will be overcome
- details of new products and estimated dates of introduction

- profitability of each product
- details of research and development projects
- after-sales support.

Markets and marketing strategy

This section of the business plan should show that thought has been given to the potential of the product. In this regard it can often be useful to identify major competitors and make an overall assessment of their strengths and weaknesses, including the following:

- an overall assessment of the market, setting out its size and growth potential
- a statement showing your position within the market
- an identification of main customers and how they compare
- details of typical orders and buying habits
- pricing strategy
- anticipated effect on demand of pricing
- expectation of price movement
- details of promotions and advertising campaigns.

It is important to identify your customers and why they might buy from you. Those entering the domestic side of the business will need to think about the best way to reach potential customers. Are local word-of-mouth recommendations enough to provide reasonable work continuity. If not, what is the most effective method of advertising to reach your customer base?

Remember, advertising is costly. It is a waste of funds to place an advertisement in a paper circulating in areas A, B, C & D if the business only covers area A.

Research and development

If you are developing a product or a particular service, then an assessment should be made on what stage it is at and what further finance is required to complete it. It may also be useful to make an assessment on the vulnerability of the product or service to innovations being initiated by others.

Basis of operation

Detail what facilities you will require in order to carry on your trade in the form of property, working and storage areas, office space, etc. An assessment should also he made on the assistance you will require from others. Your business plan might include:

- a layman's guide to the process or work
- details of facilities, buildings and plant
- key factors affecting production, such as yields and wastage
- raw material demand and usage.

Management

This section is one of the most important because it demonstrates the capability of the would-be businessman. The skills you need will cover production, marketing, finance and administration. In the early stages you may be able to do this yourself but as the business grows it may be required to develop a team to handle these matters. The following points should be considered for inclusion in the plan:

- set out age, experience and achievements
- state additional management requirements in the future and how they are to be met
- identify current weaknesses and how they will be overcome
- state remuneration packages and profit expectations
- give detailed CVs in appendices.

Advertising and retraining may be required in order to identify and provide suitable personnel where expertise and experience are lacking.

Financial information

It is important to detail, if any, the present financial position of your business and the budgeted profit and loss accounts, cash flows and balance sheets. These integrated forecasts should be prepared for the next twelve months at monthly intervals and annually for the following two years.

If the forecasts are to be reasonably accurate then the businessman must make some early decisions about:

- the premises where the business will be based, the initial repairs and alterations that might he required and an assessment of the total cost
- which plant, equipment and transport are needed, whether they are to be leased or purchased and what the cost will be?
- how much stock of materials, if any, should be carried? - the bare minimum only should be acquired, so reliable suppliers should be found
- what will be the weekly bills for overheads, wages and the proprietor's living costs?
- what type of work is going to be undertaken, and how much profit can realistically be obtained?
- how often are invoices to be presented?

Your business plan should include the following information:

- explanation of how sales forecasts are prepared
- levels of production
- details of major variable overheads and estimates
- assumptions in cash flow forecasting, inflation and taxation.

Finance required and its application

The financial details given above should produce an accurate assessment of the funds required to finance the business. It is important to distinguish between those items that require permanent finance and those that will eventually be converted to cash because it is not usually advisable to finance long-term assets with personal equity.

Working capital such as stock and debtors can usually be obtained by an overdraft arrangement but your accountant or bank will advise you on this.

Executive summary

Although it is prepared last, this summary will be the first part of your business plan. Remember that business plans are prepared for busy people and their decision on finance may be based solely on this section. It should cover two or three pages and deal with the most important aspects and opportunities in your plan. Here are some of the main headings:

- key strategies
- finance required and how it is to be used
- management experience
- anticipated returns and profits
- markets.

The appendices should include:

- CVs of key personnel
- organisation charts
- market studies
- product advertising literature
- professional references
- financial forecasts
- glossary of terms.

If you feel that any additional information should be provided in support of your proposal, then this is usually best included in the appendices.

Follow up

Please remember that once your plan is prepared, it is important to re-- examine it regularly and update the forecasts and financial information. This is a working document and can be an important tool in running the business.

Sources of finance

Personal funds

Finance, like charity, often begins at home and a would-be businessman should make a realistic assessment of his net worth, including the value of his house after deducting the mortgage(s) outstanding on it, savings, any car or van owned and any sums which the family are prepared to contribute but deducting any private borrowings which will come due for payment. The whole of these funds may not be available (for instance, money which has been loaned to a friend or relative who is known to be unable to repay at the present time).

It may not be desirable that all capital should be put at risk on a business venture so the following should be established:

- how much cash you propose to invest in the business
- whether the family home will be made available for any business borrowing
- state total finance required
- how finance is anticipated being raised
- interest and security to be provided
- expected return on investment.

Whilst it may be wise not to pledge too much of the family assets, it has to be remembered that the bank will be looking closely at the degree to which the proprietor has committed himself to the venture and will not be impressed by an application for a loan where the applicant is prepared to risk only a small fraction of his own resources.

Having decided how much of his own funds to contribute, the businessman can now see the level of shortfall and consider how best to fill it. Consideration should be given to partners where the shortfall is large and particularly when there is a need for heavy investment in fixed assets, such as premises and capital equipment. It may be worthwhile starting a limited company with others also subscribing capital and to allow the banks to take security against the book debts.

Banks

The first outside source of money to which most businessmen turn is the bank and here are a few guidelines on approaching a bank manager:

- present your business plan to him; remember to use conservative estimates which tend to understate rather than overstate the forecast sales and profits
- know the figures in detail and do not leave it to your accountant to explain them for you. The bank manager is interested in the businessman not his advisers and will be impressed if the businessman demonstrates a grasp of the financing of his business

- understand the difference between short- and long-term borrowing
- ask about the Government Loan Guarantee Scheme if there is a shortage of security for loans. The bank may be able to assist, or depending on certain conditions being met, the Government may guarantee a certain percentage of the bank loan.

Remember the bank will want their money back, so bank borrowings are usually required to be secured by charges on business assets. In start-up situations, personal guarantees from the proprietors are normally required. Ensure that if these are given they are regularly reviewed to see if they are still required.

Enterprise Investment Scheme - business angels

If an outside investor is sought in a business he will probably wish to invest within the terms of the Enterprise Investment Scheme which enables him to gain income tax relief at 20% on the amount of his investment. Additionally, any investment can be used to defer capital gains tax. The rules are complex and professional advice should always be sought.

Hire purchase/leasing

It is not always necessary to purchase assets outright that are required for the business and leasing and hire purchase can often form an integral part of a business's medium-term finance strategy.

Venture capital

In addition, there are a number of other financial institutions in the venture capital market that can help well-established businesses, usually limited companies, who wish to expand. They may also assist well-conceived start-ups. They will provide a flexible package of equity and loan capital but only for large amounts, usually sums in excess of £150,000 and often £250,000.

Usually the deal involves the financial institution having a minority interest in the voting share capital and a seat on the board of the company. Arrangements for the eventual purchase of the shares held by the finance company by the private shareholders are also normally incorporated in the scheme.

The Royal Jubilee and Princes Trust

These trusts through the Youth Business Initiative provide bursaries of not more than £1,000 per individual to selected applicants who are unemployed and age 25 or over. Grants may be used for tools and equipment, transport, fees, insurance, instruction and training but not for working capital, rent and rates, new materials or stock. They operate through a local representative whose name and address may be ascertained by contacting the Prince's Youth Business.
Point of contact: telephone 0207-321 6500.

The Business Start-up Scheme

This is an allowance of £50 per week, in addition to any income made from your business, paid for twenty weeks. To qualify you must be at least 18 and under 65, work at least 36 hours per week in the business and have been unemployed for at least six months or fall into one of the other categories: disabled, ex-HMS or redundant.

The first step is to get the booklet on the subject from your local Jobcentre or TEC that includes details on how and where to apply. Once in receipt of the enterprise allowance, you will also have the benefit of advice and assistance from an experienced businessman from your TEC. All the initial counselling services and training courses are free.

Running a business

Many businesses are run without adequate information being available to check trend in their vital areas, e.g. marketing, money and managerial efficiency. It is essential to look critically at all aspects of the business in order to maximise profits and reduce inefficiency. Regular meaningful information is required on which management can concentrate. This will vary according to the proprietor's business but will often concentrate on debtors, creditors, cash, sales and orders.

Proprietors often have the feeling that the business should be 'doing better' but are unable to identify what is going wrong. Sometimes there is the worrying phenomenon of a steadily increasing work programme coupled with a persistently reducing bank balance or rising overdraft. Some useful ways of checking the position and of identifying problem areas are given below.

Marketing

Throughout his business life the entrepreneur should continuously study the methods and approach of his competitors. A shortcoming frequently found in ailing concerns is that the proprietor thinks he knows what his customers want better than they do.

The term 'market research' sounds both difficult and expensive but a very simple form of it can be done quite effectively by the businessman and his sales staff. Existing and prospective customers should be approached and asked what they want in terms of price, quality, design, payment terms, follow-up service, guarantees and services.

The initial approach might be by a leaflet or letter followed by a personal call. As an on-going part of management, all staff with customer contact should be encouraged to enquire about and record customer preferences, complaints, etc. and feed it back to management.

Other sources of information can be trade and business journals, trade exhibitions, suppliers and representatives from which information about trends, new techniques and products can be obtained and studied. Valuable information can also be gained from studying competitors and the following questions should be asked:

- what do they sell and at what prices?
- what inducements do they offer to their customers, e.g. credit facilities, guarantees, free offers and discounts?
- how do they reach their customers - local/national advertising, mail shots, salesmen, local radio and TV?
- what are the strongest aspects of their appeal to customers and have they any weaknesses?

The businessman should apply all the information gathered from customers and competitors to his own services with a view to making sure he is offering the right product at the right price in the most attractive way and in the most receptive market.

In a small business where the proprietor is also his own salesman he must give careful thought on how he can best present his product and himself. For instance, if he is working solely within the construction industry his main problems are likely to centre on getting a C1S6 Certificate and using trade contacts to get sub-contract work.

However, for those who serve the general public, presentation can be a vital element in getting work. The customer is looking for efficiency, reliability and honesty in a trader and quality, price and style in the product. To bring out these facets in discussion with a potential customer is a skilled task. A short course on marketing techniques could pay handsome dividends. The Business Link will give the names and addresses of such courses locally.

Financial control

Unfortunately, some unsuccessful firms do not seek financial advice until too late when the downward trend cannot be halted. Earlier attention to the problems may have saved some of them so it is important to recognise the tell-tale signs. There are some tests and checks that can be done quite easily.

Cash flow

Cash flow is the lifeblood of the business and more businesses fail through lack of cash than for any other reason. Cash is generated through the conversion of work into debtors and then into payment and also through the deferral of the payment of supplies for as long a period that can be negotiated. The objective must be to keep stock, work in progress, debts to a minimum and creditors to a maximum.

Debtor days

This is calculated by dividing your trade debtors by annual sales and multiplying by 365. This shows the number of days' credit being afforded to your customers and should be compared both with your normal trade terms and the previous month's figures. Normal procedures should involve the preparation of a monthly-aged list of debtors showing the name of the customer, the value and to which month it relates.

The oldest and largest debtors can be seen at a glance for immediate consideration of what further recovery action is needed. The list may also show over-reliance on one or two large customers or the need to stop supplying a particularly bad payer until his arrears have been reduced to an acceptable level. Consideration should be given to making up bills to a date before the end of the month and making sure the accounts are sent out immediately, followed by a statement four weeks later.

Consider giving discounts for prompt payment. If all else fails, and legal action for recovery is being contemplated, call at the County Court and ask for their leaflets.

Stock turn

The level of stock should be kept to a minimum and the number of days' stock can be calculated by dividing the stock by the annual purchases and multiplying by 365. A worsening trend on a month-by-month basis shows the need for action. It is important to regularly make a full inventory of all stock and dispose of old or surplus items for cash. A stock control procedure to avoid stock losses and to keep stock to a minimum should be implemented.

Profitability

Whilst cash is vital in the short-term, profitability is vital in the medium-term. The two key percentage figures are the gross profit percentage and the net profit percentage. Gross profit is calculated by deducting the cost of materials and direct labour from the sales figures whilst net profit is arrived at after deducting all overheads. Possible reasons for changes in the gross profit percentage are:

- not taking full account of increases in materials and wages in the pricing of jobs
- too generous discount terms being offered
- poor management, over-manning, waste and pilferage of materials
- too much down-time on equipment which is in need of replacement.

If net profit is deteriorating after the deduction of an appropriate reward for your own efforts, including an amount for your own personal tax liability, you should review each item of overhead expenditure in detail asking the following questions:

- can savings be made in non-productive staff?
- is sub-contracting possible and would it be cheaper?
- have all possible energy-saving methods been fully explored?
- do the company's vehicles spend too much time in the yard and can they be shared or their number reduced?
- is the expenditure on advertising producing sales - review in association with 'marketing' above?

Over-trading

Many inexperienced businessmen imagine that profitability equals money in the bank and in some cases, particularly where the receipts are wholly in cash, this may be the case. But often, increased business means higher stock inventories, extra wages and overheads, increased capital expenditure on premises and plant, all of which require short-term finance.

Additionally, if the debtors show a marked increase as the turnover rises, the proprietor may find to his surprise that each expansion of trade reduces rather than increases his cash resources and he is continually having to rely on extensions to his existing credit.

The business, which had enough funds for start-up, finds it does not have sufficient cash to run at the higher level of operation and the bank manager may he getting anxious about the increasing overdraft. It is essential for those who run a business that operates on credit terms to be aware that profitability does not necessarily mean increased cash availability. Regular monthly management information on marketing and finance as described in this chapter will enable over-trading to be recognised and remedial action to be taken early.

If the situation is appreciated only when the bank and other creditors are pressing for money, radical solutions may be necessary, such as bringing in new finance, sale and leaseback of premises, a fundamental change in the terms of trade or even selling out to a buyer with more resources. Help from the firm's accountant will be needed in these circumstances.

Break-even point

The costs of a business may be divided into two types - variable and fixed. *Variable costs* are those which increase or decrease as the volume of work goes up or down and include such items as materials used, direct labour and power machine tools. *Fixed costs* are not related to turnover and are sometimes called fixed overheads. They include rent, rates, insurance, heat and light, office salaries and plant depreciation. These costs are still incurred even though few or no sales are being made.

Many small businessmen run their enterprises from home using family labour as back-up; they mainly sell their own labour and buy materials and hire plant only as required. By these means they reduce their fixed costs to a minimum and start making profits almost immediately. However, larger firms that have business premises, perhaps a small workshop, an office and vehicles, need to know how much they have to sell to cover their costs and become profitable.

In the case of a new business it is necessary to estimate this figure but where annual accounts are available a break-even chart based on them can be readily prepared. Suppose the real or estimated figures (expressed in £000s) are:

	%	£
Sales	100	400
Variable costs	66	265
Gross profit	34	135
Fixed costs	13	50
Net profit	21	85

Break-even point = $\frac{\text{50 divided by (1 less variable costs \%)}}{\text{sales}}$

= 50 divided by (1 less 0.6625)
= 50 divided by 0.3375
= £148 (thousand)

In practice, things are never quite as clear cut as the figures show, but nevertheless this is a very useful tool for assessing not only the break-even point but also the approximate amount of loss or profit arising at differing levels of turnover and also for considering pricing policy.

Taxation

The first decision usually required to be made from a taxation point of view is which trading entity to adopt. The options available are set out below.

Sole trader

A sole trader is a person who is in business on his own account. There is no statutory requirement to produce accounts nor is there a necessity to have them audited. A sole trader may, however, be required to register for PAYE and VAT purposes and maintain records so that Income Tax and VAT returns can be made. A sole trader is personally liable for all the liabilities of his business.

Partnership

A partnership is a collection of individuals in business on their own account and whose constitution is generally governed by the Partnership Act 1890. It is strongly recommended that a partnership agreement is also established to determine the commercial relationship between the individuals concerned.

The requirements in relation to accounting records and returns are similar to those of a sole trader and in general a partner's liability is unlimited.

Limited company

This is the most common business entity. Companies are incorporated under the Companies Act 1985 which requires that an annual audit is carried out for all companies with a turnover in excess of £5,000,000 or a review if the turnover is less than £5,000,000 and that accounts are filed with the Companies Registrar. Generally an individual shareholder's liability is limited to the amount of the share capital he is required to subscribe.

Advantages

In view of the problems and costs of incorporating an existing business, it is important to try and select the correct trading medium at the commencement of operations. It is not true to say that every business should start life as a company.

Many businesses are carried on in a safe and efficient manner by sole traders or partnerships. Whilst recognising the possible commercial advantages of a limited company, taxation advantages exist for sole traderships and partnerships, such as income tax deferral and National Insurance saving. No decision should be taken without first seeking professional advice.

The benefit of limited liability should not be ignored although this can largely be negated by banks seeking personal guarantees. In addition, it may be easier for the companies to raise finance because the bank can take security on the debts of the company that could be sold in the future, particularly if third-party finance has been obtained in the form of equity.

Self-assessment

From the tax year 1996/97 the burden of assessing tax shifted from the Inland Revenue to the individual tax payer. The main features of this system are as follows:

- the onus is on the taxpayer to provide information and to complete returns
- tax will be payable on different dates
- the taxpayer has a choice: he can calculate his tax liability at the same time as making his return and this will need to be done by 31st January following the end of the tax year. Alternatively, he can send in his tax return before 30 September and the Inland Revenue will calculate the tax to be paid on the following 31 January
- the important aspect to the system is that if the return is late, or the tax is paid late, there will be automatic penalties and/or surcharges imposed on the taxpayer.

Tax correspondence

Businessmen do not like letters from the Inland Revenue but they should resist the temptation to tear them up or put them behind the clock and forget about them. All Tax Calculations and Statements of Account should be checked for accuracy immediately and any queries should be put to your accountant or sent to the Tax District that issued the document.

Keep copies of all correspondence with the Inland Revenue. Letters can be mislaid or fail to be delivered and it is essential to have both proof of what was sent as well as a permanent record of all correspondence.

Dates tax due

Income Tax

Payments on account (based on one half of last year's liability) are due on 31 January and 31 July. If these are insufficient there is a balancing payment due on the following 31 January – the same day as the tax return needs to be filed. For example:

for the year 2004/05	Tax due £5,000 (2003/04 was £4,000)
	First payment on account of £2,000 is due on 31.01.05
	Second payment on account of £2,000 is due on 31.07.05
	Balancing payment of £1,000 is due on 31.01.06

Note that on 31.01.06, the first payment on account of £2,500 fell due for the tax year 2005/06.

Tax in business

Spouses in business

If spouses work in the business, perhaps answering the phone, making appointments, writing business letters, making up bills and keeping the books, they should be properly remunerated for it. Being a payment to a family member, the Inspector of Taxes will be understandably cautious in allowing remuneration in full as a business expense. The payment should be:

- actually paid to them, preferably weekly or monthly and in addition to any housekeeping monies
- recorded in the business book

- reasonable in amount in line with their duties and the time spent on them.

If the wages paid to them exceed £91.00 per week, Class 1 employer's and employee's NIC becomes due and if they exceed £4,745 p.a. (assuming they have no other income) PAYE tax will also be payable.

It should also be noted that once small businesses are well established and the spouses' earnings are approaching the above limits, consideration may be given to bringing them in as a partner. This has a number of effects:

- there is a reduced need to relate the spouse's income (which is now a share of the profits) to the work they do
- they will pay Class 2 and Class 4 NIC instead of the more costly Class I contributions and PAYE will no longer apply to their earnings but remember that, as partners, they have unlimited liability.

Premises

Many small businessmen cannot afford to rent or buy commercial premises and run their enterprises from home using part of it as an office where the books and vouchers, clients' records and trade manuals are kept and where estimates and plans are drawn up. In these circumstances, a portion of the outgoings on the property may be claimed as a business expenses. An accountant's advice should be sought to ensure that the capital gains tax exemption that applies on the sale of the main residence is not lost.

Fixed Profit Car Scheme

It may be advantageous to calculate your car expenses using a fixed rate per business mile. A condition is that your annual turnover is below the VAT threshold (currently £58,000). Ask your accountant about this. A proper record of business mileage must be kept.

Vehicles

Car expenses for sole traders and partners are usually split on a fractional mileage basis between business journeys, which are allowable, and private ones, which are not, and a record of each should he kept. If the business does work only on one or two sites for only one main contractor, the inspector may argue that the true base of operations is the work site not the residence and seek to disallow the cost of travel between home and work. It is tax-wise and sound business practice to have as many customers as possible and not work for just one client.

Business entertainment

No tax relief is due for expenditure on business entertainment and neither is the VAT recoverable on gifts to customers, whether they are from this country or overseas. However, the cost of small trade gifts not exceeding £50 per person per annum in value is still admissible provided that the gift advertises the business and does not consist of food, drink or tobacco.

Income tax (2004/05)

Personal allowances

The current personal allowance for a single person is £4,745. The personal allowance for people aged 65 to 74 and over 75 years are £6,830 and £6,950 respectively. The married couple's allowance was withdrawn on 5 April 2000, except for those over 65 on that date.

Taxation of husband and wife

A married woman is treated in much the same way as a single person with her own personal allowance and basic rate band. Husband and wife each make a separate return of their own income and the Inland Revenue deals with each one in complete privacy; letters about the husband's affairs will be addressed only to him and about the wife's only to her unless the parties indicate differently.

Rates of tax

Tax is deducted at source from most banks and building societies accounts at the rate of 20%. The rates of tax for 2004/05 are as follows:

Lower rate: 10% on taxable income up to £2,020
Basic rate: 22% on taxable income between £2,020 and £31,400
Higher rate: 40% on taxable income over £31,400

Dividends carry a 10% non-repayable tax credit. Higher rate taxpayers pay a further tax on dividends of 22.5%.

Mortgage interest relief

This is no longer available after 5 April 2000.

Business losses

These are allowed only against the income of the person who incurs the loss. For example, a loss in the husband's business cannot be set against the wife's income from employment.

Joint income

In the case of joint ownership by a husband and wife of assets that yield income, such as bank and building society accounts, shares and rented property, the Inland Revenue will treat the income as arising equally to both and each will pay tax on one half of the income. If, however, the asset is owned in unequal shares or one spouse only and the taxpayer can prove this, then the shares of income to be taxed can be adjusted accordingly if a joint declaration is made to the tax office setting out the facts.

Capital Gains Tax

Where an asset is disposed of, the first £8,200 of the gain is exempt from tax. In the case of husbands and wives, each has a £8,200 exemption so if the ownership of the assets is divided between them, it is possible to claim exemption on gains up to £16,400 jointly in the tax year. Any remaining gain is chargeable as though it were the top slice of the individual's income; therefore according to his or her circumstances it might be charged at 10%, 22% or 40%.

Self-employed NIC rates (from 6 April 2004)

Class 2 rate
Charged at £2.00 per week. If earnings are below £4,745 per annum averaged over the year, ask the DSS about 'small income exception'. Details are in leaflet CA02.

Class 4 rate
Business profits up to £4,745 per annum are charged at NIL. Annual profits between £4,615 and £31,720 are charged at 8% of the profit. There is also a charge on profits over £31,720 of 1%. Class 4 contributions are collected by the Inland Revenue along with the income tax due.

Capital allowances (depreciation) rates

Plant and machinery:	25% (40% first-year allowance is available for certain small businesses)
Business motor cars - cost up to £12,000:	25%
- cost over £12,000:	£3,000 (maximum)
Industrial build	4%
Commercial and industrial buildings in Enterprise Zones:	100%
Computers and software equipment	50%

THE CONSTRUCTION INDUSTRY TAX DEDUCTION SCHEME

General

The new Construction Industry Tax Deduction Scheme is known as the 'CIS' scheme and replaced the old '714' scheme. As the scheme operates whenever a contractor makes a payment to a sub-contractor, the businessman should visit his local income tax enquiry office and obtain copies of the Inland Revenue booklet IR 14/15 (CIS) and leaflet IR 40 which will explain the conditions under which the Inland Revenue will issue a registration card or (CIS6) certificate and precisely when the scheme applies.

Everyone who carries out work in the Construction Industry Scheme must hold a registration card (CIS4) or a tax certificate (CIS6). Certain larger companies use a special certificate (CIS5).

If the sub-contractor has a registration card but does not hold a valid tax certificate (CIS6) issued to him by the Inland Revenue, then the contractor *must* deduct 18% tax from the whole of any payment made to him (excluding the cost of any materials) and to account to the Inland Revenue for all amounts so withheld.

To enable the subcontractor to prove to the Inspector of Taxes that he has suffered this tax deduction, the contractor must complete the three-part tax payment voucher (CIS25) showing the amount withheld. These vouchers must be carefully filed for production to the Inspector after the end of the tax year along with the tax return. Any tax deducted in this way over and above the sub-contractor's agreed liability for the year will be repaid by the Inland Revenue. If he holds a (CIS6) certificate the payment may be made in full without deducting tax.

A small business that does work only for the general public and small commercial concerns is outside the scheme and does not need a certificate to trade. If, however, it engages other contractors to do jobs for it, the business would have to register under the scheme as a contractor and deduct tax from any payment made to a sub-contractor who did not produce a valid (CIS6) certificate. If in doubt, consult your accountant or the Inland Revenue direct.

VAT

The general rule about liability to register for VAT is given in the VAT office notes. It is possible to give here only a brief outline of how the tax works. The rules that apply to the construction industry are extremely complex and all traders must study *The VAT Guide* and other publications.

Registration for VAT is required if, at the end of any month, the value of taxable supplies in the last 12 months exceeds the annual threshold or if there are reasonable grounds for believing the value of the taxable supplies in the next 30 days will exceed the annual threshold.

Taxable supplies include any zero-rated items. The annual threshold is £58,000. The amount of tax to be paid is the difference between the VAT charged out to customers *(output tax)* and that suffered on payments made to suppliers for goods and services *(input tax)* incurred in making taxable supplies. Unlike income tax there is no distinction in VAT for capital items so that the tax charged on the purchase of, for example, machinery, trucks and office furniture, will normally be reclaimable as *input tax*.

VAT is payable in respect of three monthly periods known as 'tax periods'. You can apply to have the group of tax periods that fits in best with your financial year. The tax must be paid within one month of the end of each tax period. Traders who receive regular repayments of VAT can apply to have them monthly rather than quarterly. Not all types of goods and services are taxed at 17.5% (i.e. the standard rate). Some are exempt and others are zero-rated.

Zero-rated

This means that no VAT is chargeable on the goods or services, but a registered trader can reclaim any *input* tax suffered on his purchases. For instance, a builder pays VAT on the materials he buys to provide supplies of constructing but if he is constructing a new dwelling house, this is zero rated. The builder may reclaim this VAT or set it off against any VAT due on standard rated work.

Exempt

Supplies that are exempt are less favourably treated than those that are zero rated. Again no VAT is chargeable on the goods or services but the trader cannot reclaim any *input* tax suffered on his purchases.

Standard-rated

All work which is not specifically stated to be zero rated or exempt is standard-rated, i.e. VAT is chargeable at the current rate of 17.5% and the trader may deduct any *input* tax suffered when he is making his return to the Customs and Excise. If for any reason a trader makes a supply and fails to charge VAT when he should have done so (e.g. mistakenly assuming the supply to be zero rated), he will have to account for the VAT himself out of the proceeds. If there is any doubt about the VAT position, it is safer to assume the supply is standard rated, charge the appropriate amount of VAT on the invoice and argue about it later.

Time of supply

The *time* at which a supply of goods or services is treated as taking place is important and is called the 'tax point'. VAT must be accounted for to the Customs and Excise at the end of the accounting period in which this 'tax point' occurs. For the supply of goods which are 'built on site', the 'basic tax point' is the date the goods are made available for the customer's use, whilst for *services* it is normally the date when all work except invoicing is completed.

However, if you issue a tax invoice or receive a payment before this 'basic tax point' then that date becomes a tax point. In the case of contracts providing for stage and retention payments, the tax point is either the date the tax invoice is issued or when payment is received, whichever is the earlier.

All the requirements apply to sub-contractors and main contractors and it should be noted that, when a contractor deducts income tax from a payment to a sub-contractor (because he has no valid CIS6) VAT is payable on the full gross amount *before* taking off the income tax.

Annual accounting

It is possible to account for VAT other than on a specified three month period. Annual accounting provides for nine equal installments to be paid by direct debit with annual return provided with the tenth payment. £300,000.

Cash accounting

If turnover is below a specified limit, currently £660,000, a taxpayer may account for VAT on the basis of cash paid and received. The main advantages are automatic bad debt relief and a deferral of VAT payment where extended credit is given.

Bad debts

Relief is available for debts over 6 months.

Part Six

GENERAL CONSTRUCTION DATA

GENERAL CONSTRUCTION DATA

The metric system

Linear

1 centimetre (cm)	=	10 millimetres (mm)
1 decimetre (dm)	=	10 centimetres (cm)
1 metre (m)	=	10 decimetres (dm)
1 kilometre (km)	=	1000 metres (m)

Area

100 sq millimetres	=	1 sq centimetre
100 sq centimetres	=	1 sq decimetre
100 sq decimetres	=	1 sq metre
1000 sq metres	=	1 hectare

Capacity

1 millilitre (ml)	=	1 cubic centimetre (cm3)
1 centilitre (cl)	=	10 millilitres (ml)
1 decilitre (dl)	=	10 centilitres (cl)
1 litre (l)	=	10 decilitres (dl)

Weight

1 centigram (cg)	=	10 milligrams (mg)
1 decigram (dg)	=	10 centigrams (mcg)
1 gram (g)	=	10 decigrams (dg)
1 decagram (dag)	=	10 grams (g)
1 hectogram (hg)	=	10 decagrams (dag)

Conversion equivalents (imperial/metric)

Length

1 inch	=	25.4 mm
1 foot	=	304.8 mm
1 yard	=	914.4 mm
1 yard	=	0.9144 m
1 mile	=	1609.34 m

Area

1 sq inch	=	645.16 sq mm
1 sq ft	=	0.092903 sq m
1 sq yard	=	0.8361 sq m
1 acre	=	4840 sq yards
1 acre	=	2.471 hectares

Liquid

1 lb water	=	0.454 litres
1 pint	=	0.568 litres
1 gallon	=	4.546 litres

Horse-power

1 hp	=	746 watts
1 hp	=	0.746 kW
1 hp	=	33,000 ft.lb/min

Weight

1 lb	=	0.4536 kg
1 cwt	=	50.8 kg
1 ton	=	1016.1 kg

Conversion equivalents (metric/imperial)

Length

1 mm	=	0.03937 inches
1 centimetre	=	0.3937 inches
1 metre	=	1.094 yards
1 metre	=	3.282 ft
1 kilometre	=	0.621373 miles

Area

1 sq millimetre	=	0.00155 sq in
1 sq metre	=	10.764 sq ft
1 sq metre	=	1.196 sq yards
1 acre	=	4046.86 sq m
1 hectare	=	0.404686 acres

Weight

1 kg	=	2.205 lbs
1 kg	=	0.01968 cwt
1 kg	=	0.000984 ton

Temperature equivalents

In order to convert Fahrenheit to Celsius deduct 32 and multiply by 5/9. To convert Celsius to Fahrenheit multiply by 9/5 and add 32.

Fahrenheit	Celsius
230	110.0
220	104.4
210	98.9
200	93.3
190	87.8
180	82.2
170	76.7
160	71.1
150	65.6
140	60.0
130	54.4
120	48.9
110	43.3
100	37.8
90	32.2
80	26.7
70	21.1
60	15.6
50	10.0
40	4.4
30	-1.1
20	-6.7
10	-12.2
0	-17.8

Areas and volumes

Figure	Area	Perimeter
Rectangle	Length × breadth	Sum of sides
Triangle	Base × half of perpendicular height	Sum of sides
Quadrilateral	Sum of areas of contained triangles	Sum of sides
Trapezoidal	Sum of areas of contained triangles	Sum of sides
Trapezium	Half of sum of parallel sides × perpendicular height	Sum of sides
Parallelogram	Base × perpendicular height	Sum of sides
Regular polygon	Half sum of sides × half internal diameter	Sum of sides
Circle	pi × radius^2	pi × diameter or pi × 2 × radius

Figure	Surface area	Volume
Cylinder	pi × 2 × radius^2 × length (curved surface only)	pi × radius^2 × length
Sphere	pi × diameter^2	Diameter^3 × 0.5236

Weights of materials	kg/m3	kg/m^2	kg/m
Aggregate, coarse	1,500		
Ashes	800		
Ballast	600		

Weights of materials	**kg/m3**	**kg/m²**	**kg/m**
Blocks, natural aggregate			
75mm		160	
100mm		215	
140mm		300	
Blocks, lighweight aggregate			
75mm		60	
100mm		80	
140mm		112	
Bricks, Fletton		1820	
Bricks, engineering		2250	
Bricks, concrete		1850	
Brickwork, 112.5mm		220	
Brickwork, 215mm		465	
Brickwork, 327.5mm		710	
Cement	1,440		
Chalk	2,240		
Flint	2,550		
Gravel	1,750		
Hardcore	1,900		
Hoggin	1,750		
Lime, ground	750		
Stone, natural	2,400		
Stone, Portland	2,200		
Stone, reconstructed	2,250		
Stone, York	2,400		

EXCAVATION AND FILLING

Shrinkage of deposited material

Clay	-10%
Gravel	-7.50%
Sandy soil	-12.50%

Bulking excavated material

Clay	40%
Gravel	25%
Sand	20%

Typical fuel consumption for plant	**Engine size kW**	**Litres per hour**
Compressors up to	20.00	4.00
	30.00	6.50
	40.00	8.20
	50.00	9.00
	75.00	16.00
	100.00	20.00
	125.00	25.00
	150.00	30.00
Concrete mixers up to	5.00	1.00
	10.00	2.40
	15.00	3.80
	20.00	5.00
	20.00	5.00
Dumpers	5.00	1.30
	7.00	2.00
	10.00	3.00
	15.00	4.00
	20.00	4.90
	30.00	7.00
	50.00	12.00
Excavators	10.00	2.50
	20.00	4.50
	40.00	9.00
	60.00	13.00
	80.00	17.00

	Engine size kW	Litres per hour
Pumps	5.00	1.10
	10.00	2.10
	15.00	3.20
	20.00	4.20
	25.00	5.50

CONCRETE WORK

Concrete mixes	Cement t	Sand m3	Aggregate m3	Water litres
1:1:2	0.50	0.45	0.70	208.00
1:1:5:3	0.37	0.50	0.80	185.00
1:2:4	0.30	0.54	0.85	175.00
1:3:6	0.22	0.55	0.85	160.00

BRICKWORK AND BLOCKWORK

Bricks per m2 (brick size 215 × 103.5 × 65mm)	nr
Half brick wall	
stretcher bond	59
English bond	89
English garden wall bond	74
Flemish bond	79
One brick wall	
English bond	118
Flemish bond	118

Bricks per m2 (cont'd)	**nr**
One and a half brick wall	
English bond	178
Flemish bond	178
Two brick wall	
English bond	238
Flemish bond	238
Metric modular bricks	
200 × 100 × 75mm	
90mm thick	133
190mm thick	200
200 × 100 × 100mm	
90mm thick	50
190mm thick	100
290mm thick	150
300 × 100 × 75mm	
90mm thick	44
300 × 100 × 100mm	
90mm thick	50

Blocks per m2 (block size 414 × 215mm)

60mm thick	9.9
75mm thick	9.9
100mm thick	9.9
140mm thick	9.9
190mm thick	9.9
215mm thick	9.9

Mortar per m2	**Wirecut m3**	**1 Frog m3**	**2 Frogs m3**
Brick size 215 × 103.5 × 65mm			
Half brick wall	0.017	0.024	0.031
One brick wall	0.045	0.059	0.073
One and a half brick wall	0.072	0.093	0.114
Two brick wall	0.101	0.128	0.155

Brick size 200 × 100 × 75mm	**Solid m3**	**Perforated m3**
90mm thick	0.016	0.019
190mm thick	0.042	0.048
290mm thick	0.068	0.078
Brick size 200 × 100 × 100mm		
90mm thick	0.013	0.016
190mm thick	0.036	0.041
290mm thick	0.059	0.067
Brick size 200 × 100 × 100mm		
90mm thick	0.015	0.018
Block size 440 × 215mm		
60mm thick	0.004	
75mm thick	0.005	
100mm thick	0.006	
140mm thick	0.007	
190mm thick	0.008	
215mm thick	0.009	

MASONRY

Mortar per m2 of random rubble walling	m3
300mm thick wall	0.120
450mm thick wall	0.160
550mm thick wall	0.120

EXTERNAL WORKS

Blocks/slabs per m2	nr/m2
200 × 100mm	50.00
450 × 450mm	4.93
600 × 450mm	3.70
600 × 600mm	2.79
600 × 750mm	2.22
600 × 900mm	1.85

Drainage trench widths	Under 1.5m deep mm	Over 1.5m deep mm
Pipe diameter 100mm	450	600
Pipe diameter 150mm	500	650
Pipe diameter 225mm	600	750
Pipe diameter 300mm	650	800

Volumes of filling for pipe beds (m3 per m)	50mm thick	100mm thick	150mm thick
Pipe diameter 100mm	0.023	0.045	0.068
Pipe diameter 150mm	0.026	0.053	0.079
Pipe diameter 225mm	0.030	0.060	0.090
Pipe diameter 300mm	0.038	0.075	0.113

Volumes of filling for pipe bed and haunching (m3 per m)

	m3
Pipe diameter 100mm	0.117
Pipe diameter 150mm	0.152
Pipe diameter 225mm	0.195
Pipe diameter 300mm	0.279

Volumes of filling for pipe bed and surround (m3 per m)

	m3
Pipe diameter 100mm	0.185
Pipe diameter 150mm	0.231
Pipe diameter 225mm	0.285
Pipe diameter 300mm	0.391

Part Seven

PLANT NAMES

ENGLISH/LATIN PLANT NAMES

Common/English	**Scientific/Latin**
Adam's Needle	Yucca gloriosa
Alder	Alnus
Alder Buckhorn	Frangula alnus
Algerian Iris	Iris unguicularis
Apple	Malus
Ash	Fraxinus
Atlantic Cedar	Cedrus libani atlantica
Austrian Pine	Pinus nigra austriaca
Balsam Poplar	Populus balsamifera
Bamboo	Pseudosasa japonica
Bay	Laurus nobilis
Bay Willow	Salix pentandra
Bear Tree	Arctostaphylos uva-ursi
Beech	Fagus
Birch	Betula
Bird Cherry	Prunus padus
Black Poplar	Populus nigra
Black Stemmed Bamboo	Phyllostachys nigra
Black Walnut	Juglans nigra
Blackthorn	Prunus spinosa
Bladder Nut	Staphylea colchica
Blue Holly	Ilex x meserveae
Boston Ivy	Parthenocissus tricuspidata
Box	Buxus
Bramble	Rubus
Broom	Cytisus
Broom	Cytisus scorparius
Bugle	Ajuga
Butchers Broom	Ruscus aculeatus
Butterbur	Petastites hybidus
Canary Ivy	Hedera canarienis
Cape Figwort	Phygelius capensis coccineus
Cedar	Cedrus

Common/English	**Scientific/Latin**
Cherry	Prunus
Cherry Laurel	Prunus laurocreasus
Chestnut Holly	Ilex x koeheana
Christmas Box	Sarcococca
Chusan Palm	Trachycarpus fortunei
Comfrey	Symphytum
Common Alder	Alnus glutinosa
Common Ivy	Hedera helix
Corsican Pine	Pinus nigra maritima
Crab Apple	Malus
Crack Willow	Salix fragilis
Cranesbill	Geranium
Cricket Bat Willow	Salix alba Caerulea
Cucumber Tree	Magnolia acuminata
Current	Ribes
Cypress	Cupressus
Daisy Bush	Olearia
Dawn Redwood	Metasequoia glyptostroboides
Day Lily	Hemerocallis
Dog Rose	Rosa canina
Dogwood	Cornus
Douglas Fir	Pseudotsuga menziesii
Downy Birch	Betula pubescens
Dwarf Almond	Prunus tenella
Dyers Greenweed	Genista tinctoria
Eglantine or Sweet Briar	Rosa rubiginosa
Elder	Sambucus
Elephant's Ears	Bergenia
Elisha's Tears	Leycesreria Formosa
Evergreen Magnolia	Magnolia grandiflora
Evergreen Oak	Quercus ilex
False Acacia	Robinia pseudoacacia
Field Maple	Acer campestre

Common/English	**Scientific/Latin**
Field Rose	Rosa arvenis
Fig	Ficus carica
Fir	Abies
Firethorn	Pyracantha
Flowering Ash	Fraxinus ornus
Flowering Quince	Chaenomeles
Fly Honeysuckle	Lonicera xylostenum
Foxglove Tree	Paulownia
Germander	Teucrium
Giants Reed	Arundo donax
Giant Rhubarb	Gunnera manicata
Gladwyn	Iris foetidissima
Goat Willow/Sallow	Salix caprea
Goats Beard	Aruncus dioicus
Golden Hop	Humulus lupulus aureus
Golden Oat	Stipa gigantean
Golden Rain	Koelreuteria paniculata
Gorse	Ulex
Grapevine	Vitis vinifera
Grey Alder	Alnus incana
Guelder Rose	Viburnum opulus
Handkerchief Tree	Davidia involucrate
Hard Fern	Blechnum spicant
Hawthorn	Crataegus monogyna
Hazel	Corylus avellana
Heath	Erica
Heather	Calluna vulgaris
Hedging Privet	Ligustrum ovalifolium
Hemlock	Tsuga heterophylla
Holly	Ilex
Holly	Ilex aquifolium
Honey Locust	Gleditsia
Honeysuckle	Lonicera
Hop Hornbeam	Ostrya virgini – na

Common/English	**Scientific/Latin**
Hop Tree	Ptelea trifoliate
Hornbeam	Carpinus betulus
Horse Chestnut	Aesculus hippocastanum
Hungarian Oak	Quercus frainetto
Hyssop	Hyssopus officinalis
Indian Bean Tree	Catalpa bignonioides
Indian Chestnut	Aesculus indica
Irish Yew	Taxus baccata Fastigiata
Iron Tree	Parrotia persica
Italian Alder	Alnus cordata
Italian Cypress	Cupressus sempervirens Stricta
Ivy	Hedera
Japanese Anemone	Anemone hupehensis japonic
Japanese Anemone	Anemone x hybrida
Japanese Cedar	Cryptomeria japonica
Japanese Honeysuckle	Lonicera japonica
Japanese Larch	Larix kaempferi
Japanese Maple	Acer palmatum
Jasmine	Jasminium officinale
Jerusalem Sage	Phlomis fruiticosa
Judas Tree	Cercis siliquastrum
Juniper	Juniperus communis
Kiwi Fruit	Actinida chinensis
Knotweed	Polygonum
Lad 's Love	Artemisia abrotanum
Lady Fern	Athyrium fillix-femina
Larch	Larix
Large-Leaved Lime	Tilia platyphyllos
Lauristinus	Viburnum tinus
Lavender	Lavandula
Lavender Cotton	Santolina
Lawson's Cypress	Chamaecyparis lawsoniana

Common/English	**Scientific/Latin**
Lebanese Cedar	Cedrus libani
Lenten Rose	Hellborus orientails
Leyland Cypress	Cupressocyparis leylandii
Lilac	Syringa
Lily Turf	Liriope muscari
Lime	Tilia
London Plane	Platanus acerifolia
Loquat	Eriobotrya japonica
Male Fern	Dryopteris filix-mas
Maple	Acer
Medlar	Mespilus germanica
Mexican Orange	Choisya
Mimosa	Acacia dealbata
Mock Orange	Philadelphus
Morello Cherry	Prunus avium Morello
Moss Rose	Rosa x centifolia Muscosa
Mulberry	Morus nigra
Myrtle	Myrtus communis
New Zealand Flax	Phormium
Norway Maple	Acer platanoides
Norway Spruce	Picea abies
Oak	Quercus robur
Oregon Grape	Mahonia aquifolium
Osier	Salix viminalis
Pagoda Tree	Sophora japonica
Pampas Grass	Cortaderia
Passion Flower	Passiflora caerulea
Pear	Pyrus
Periwinkle	Vinca
Persian Ivy	Hedera colchica
Pine	Pinus
Plane	Platanus

Common/English	**Scientific/Latin**
Polypody	Polypodium
Poplar	Populus
Portuguese Laurel	Prunus lusitanica
Privet	Ligustrum
Quince	Cydonia oblonga
Red Hot Poker	Kniphofla
Red Oak	Quercus rubra
Redwood	Sequoia sempervirens
Rock Rose	Helianthemum
Rosa	Rose
Rose of Sharon	Hypericum calcyinum
Rosemary	Rosmarinus
Rowan	Sorbus aucuparia
Royal Fern	Osmunda regalis
Rue	Ruta graveoleans
Sage	Salvia
Sallow	Salix cinerea
Salt Bush	Atriplex halimus
Scarlet Oak	Quercus coccinea
Scotch Rose	Rosa pimpinellilfolia
Scots Pine	Pinus sylvestris
Sea Buckhorn	Hippophae rhamnoides
Sedge	Carex
Shrubby Cinqfoil	Potentilla
Shuttlecock Fern	Matteuccia struthiopteris
Silver Birch	Betula pendula
Silver Lime	Tilia tomentosa
Silver Maple	Acer saccharinum
Small-Leaved Lime	Tilia cordata
Smoke Bush	Cotinus
Snowberry	Symphoricarpus
Snowy Mespilus	Amelanchier
Southern Beech	Nothofagus

Common/English	**Scientific/Latin**
Spanish Broom	Spartium junceum
Spanish Chestnut	Castanea sativa
Spindle	Spindle Euonymus
Spindle Tree	Euonymus europaeus
Spuce	Picea
Spurge	Euphorbia
Spurge Laurel	Rosa rubiginosa
Sycamore	Acer pseudoplatanus
Tamarisk	Tamarix
Tealeaf Willow	Salix phylicifolia
Thyme	Thymus
Tree Lupin	Lupinus arboreus
Tree Peony	Paeonia
Tree Poppy	Romneya coulteri
Tree Heaven	Ailanthus
Tulip Tree	Liriodedrun tulipifera
Turkey Oak	Quercus cerris
Vine	Vitis
Virginia Creeper	Parthenocissus quinquefolia
Violet Willow	Salix daphnoides
Walnut	Juglans regia
Wayfaring Tree	Viburnum lantana
Weeping Willow	Salix sepuilchralis Chrysocoma
Wellingtonia	Sequoiadendron giganteum
Western Red Cedar	Thuja plicata
White Mulberry	Morus alba
White Poplar	Populus alba
White Willow	Salix alba
Whitebeam	Sorbus aria
Wild Apple	Malus sylvestris
Wild Privet	Ligustrum vulgare
Wild Service Tree	Sorbus torminalis
Willow/Sallow	Salix

Common/English	Scientific/Latin
Willow	Salix reifenweide
Wing Nut	Pterocaya fraxinifolia
Winter Honeysuckle	Lonicera x purpusii
Winter Jasmine	Jasminium nudiflorum
Winterswee	Chimonanthust
Witch Hazel	Hamamelis
Wood Spurge	Euphorbia amygdaloides
Woodland Hawthorn	Crataegus oxyacantha
Woody Nightshade	Solanum dulcamara
Woody Willow	Salix lanata
Wormwood	Artemisia absinthium
Yarrow	Achillea millefolium
Yew	Taxus

LATIN/ENGLISH PLANT NAMES

Scientific/Latin	Common/English
Abies	Fir
Acacia dealbata	Mimosa
Acer	Maple
Acer campestre	Field Maple
Acer palmatum	Japanese Maple
Acer platanoides	Norway Maple
Acer pseudoplatanus	Sycamore
Acer saccharinum	Silver Maple
Acer saccharum	Sugar Maple
Achillea millefolium	Yarrow
Actinida chinensis	Kiwi Fruit
Aesculus hippocastanum	Horse Chestnut
Aesculus indica	Indian Chestnut
Ailanthus alitissima	Tree of Heaven
Ajuga	Bugle
Alnus	Alder

Scientific/Latin	**Common/English**
Alnus cordata	Italian Alder
Alnus glutinosa	Common Alder
Alnus incana	Grey Alder
Amelanchier	Snowy Mespilus
Anemone hupehensis japonica	Japanese Anemone
Anemone x hybrida	Japanese Anemone
Arbutus unedo	Strawberry Tree
Arctostaphylos uva-ursi	Bear Tree
Artemisia abrotanum	Lad‘s Love
Artemisia absinthium	Wormwood
Aruncus dioicus	Goats Beard
Arundo donax	Giants Reed
Athyrium fillix-femina	Lady Fern
Atriplex halimus	Salt Bush
Bergenia	Elephant's Ears
Betula	Birch
Betula papyrifera	Paper Birch
Betula pendula	Silver Birch
Betula pubescens	Downy Birch
Blechnum spicant	Hard Fern
Buxus	Box
Buxus sempervirens	Box
Calluna vulgaris	Heather
Carex	Sedge
Carpinus betulus	Hornbeam
Castanea sativa	Spanish Chestnut
Catalpa bignonioides	Indian Bean Tree
Cedrus	Cedar
Cedrus libani	Lebanese Cedar
Cedrus libani atlantica	Atlantic Cedar
Cercis siliquastrum	Judas Tree
Chaenomeles	Flowering Quince
Chamaecyparis lawsoniana	Lawson's Cypress
Chimonanthus	Wintersweet

Scientific/Latin	Common/English
Choisya	Mexican Orange
Cistus	Sun Rose
Cornus	Dogwood
Cortaderia	Pampas Grass
Corylus avellana	Hazel
Cotinus	Smoke Bush
Crataegus monogyna	Hawthorn
Crataegus oxyancantha	Woodland Hawthorn
Cryptomeria japonica	Japanese Cedar
Cupressocyparis leylandii	Leyland Cypress
Cupressus	Cypress
Cupressus sempervirens Stricta	Italian Cypress
Cydonia oblonga	Quince
Cytisus	Broom
Cytisus scorparius	Broom
Daphne laureola	Spurge Laurel
Davidia involucrata	Handkerchief Tree
Dryopteris filix-mas	Male Fern
Erica	Heath
Eriobotrya japonica	Loquat
Euonymus	Spindle
Euonymus europaeus	Spindle Tree
Euphorbia	Spurge
Euphorbia amygadaloides	Wood Spurge
Fagus	Beech
Ficus carica	Fig
Frangula alnus	Alder Buckhorn
Fraxinus	Ash
Fraxinus ornus	Flowering Ash
Genista tinctoria	Dyers Greenweed
Geranium	Cranesbill
Gleditsia	Honey Locust

Scientific/Latin	**Common/English**
Gunnera manicata	Giant Rhubarb
Hamamelis	Witch Haze
Hedera	Ivy
Hedera canarienis	Canary Ivy
Hedera colchica	Persian Ivy
Hedera helix	Ivy
Helianthemum	Rock Rose
Hellborus foetidus	Stinking Hellebore
Hellborus orientalis	Lenten Rose
Hemerocallis	Day Lily
Hippophae rhamnoides	Sea Buckhorn
Humulus lupulus aureus	Golden Hop
Hypericum calcyinum	Rose of Sharon
Hyssopus officinalis	Hyssop
Ilex	Holly
Ilex aquifolium	Holly
Ilex x koeheana	Chestnut Holly
Ilex x meserveae	Blue Holly
Iris foetidissima	Gladwyn
Iris unguicularis	Algerian Iris
Jasminium nudiflorum	Winter Jasmine
Jasminium officinale	Jasmine
Juglans nigra	Black Walnut
Juglans regia	Walnut
Juniperus	Juniper
Juniperus communis	Juniper
Kniphofla	Red Hot Poker
Koelreute	Golden Rain
Larix	Larch
Larix kaempferi	Japanese Larch
Laurus nobilis	Bay

Scientific/Latin	**Common/English**
Lavandula	Lavender
Leycesreria formosa	Elisha's Tears
Ligustrum	Privet
Ligustrum ovalifolium	Hedging Privet
Ligustrum vulgare	Wild Privet
Liriodedrun tulipifera	Tulip Tree
Liriope muscari	Lily Turf
Lonicera	Honeysuckle
Lonicera japonica	Japanese Honeysuckle
Lonicera x purpusii	Winter Honeysuckle
Lonicera xylostenum	Fly Honeysuckle
Lupinus arboreus	Tree Lupin
Magnolia acuminata	Cucumber Tree
Magnolia grandiflora	Evergreen Magnolia
Mahonia aquifolium	Oregon Grape
Malus	Apple
Malus sylvestris	Wild Apple
Matteuccia struthiopteris	Shuttlecock Fern
Mespilus germanica	Medlar
Metasequoia glyptostroboides	Dawn Redwood
Morus alba	White Mulberry
Morus nigra	Mulberry
Myrtus communis	Myrtle
Nothofagus	Southern Beech
Olearia	Daisy Bush
Osmunda regalis	Royal Fern
Ostrya virgini	Hop Hornbeam
Paeonia	Tree Peony
Parrotia persica	Iron Tree
Parthenocissus quinquefolia	Virgina Creeper
Parthenocissus tricuspidata	Boston Ivy
Passiflora caerulea	Passion Flower

Scientific/Latin	Common/English
Paulownia	Foxglove Tree
Petastites hybidus	Butterbur
Philadelphus	Mock Orange
Phlomis fruiticosa	Jerusalem Sage
Phormium	New Zealand Flax
Phygelius capensis coccineus	Cape Figwort
Phyllostachys nigra	Black Stemmed Bamboo
Picea	Spruce
Picea abies	Norway Spruce
Picea omorika	Serbian Spruce
Picea sitchensis	Sitka Spruce
Pinus	Pine
Pinus nigra austriaca	Austrian Pine
Pinus nigra maritime	Corsican Pine
Pinus sylvestris	Scots Pine
Platanus	Plane
Platanus acerifolia	London Plane
Polygonum	Knotweed
Polypodium	Polypody
Populus	Poplar
Populus alba	White Poplar
Populus balsamifera	Balsam Poplar
Populus nigra	Black Poplar
Potentilla	Shrubby Cinqfoil
Prunus	Cherry
Prunus avium Morello	Morello Cherry
Prunus laurocreasus	Cherry Laurel
Prunus lusitanica	Portuguese Laurel
Prunus mahaleb	St Lucia Cherry
Prunus padus	Bird Cherry
Prunus spinosa	Blackthorn
Prunus tenella	Dwarf Almond
Pseudosasa japonica	Bamboo
Pseudotsuga menziesii	Douglas Fir
Ptelea trifoliata	Hop Tree
Pterocarya fraxinifolia	Wing Nut

Scientific/Latin	Common/English
Pyracantha	Firethorn
Pyrus	Pear
Quercus	Oak
Quercus cerris	Turkey Oak
Quercus coccinea	Scarlet Oak
Quercus frainetto	Hungarian Oak
Quercus ilex	Evergreen Oak
Quercus petraea	Sessile Oak
Quercus robur	Oak
Quercus rubra	Red Oak
Rhus typhinia	Sumach
Ribes	Current
Robinia pseudoacacia	False Acacia
Romneya coulteri	Tree Poppy
Rosa	Rose
Rosa arvenis	Field Rose
Rosa canina	Dog Rose
Rosa pimpinellilfolia	Scotch Rose
Rosa rubiginosa	Eglantine or Sweet Briar
Rosa x centifolia Muscosa	Moss Rose
Rosmarinus	Rosemary
Rubus	Bramble
Ruscus aculeatus	Butchers Broom
Ruta graveoleans	Rue
Salix	Willow/Sallow
Salix alba	White Willow
Salix alba Caerulea	Cricket Bat Willow
Salix caprea	Goat Willow/Sallow
Salix cinerea	Sallow
Salix daphnoides	Violet Willow
Salix fragilis	Crack Willow
Salix lanata	Woody Willow
Salix pentandra	Bay Willow

Scientific/Latin	Common/English
Salix phylicifolia	Tealeaf Willow
Salix viminalis	Osier
Salix x sepulchralis Chrysocoma	Weeping Willow
Salvia	Sage
Sambucus	Elder
Santolina	Lavender Cotton
Sarcococca	Christmas Box
Sedum	Stonecrop
Sequoia sempervirens	Redwood
Sequoiadendron giganteum	Wellingtonia
Solanum dulcamara	Woody Nightshade
Sophora japonica	Pagoda Tree
Sorbus aria	Whitebeam
Sorbus aucuparia	Rowan
Sorbus terminalis	Wild Service Tree
Spartium junceum	Spanish Broom
Staphylea colchica	Bladder Nut
Stipa gigantea	Golden Oat
Symphoricarpus	Snowberry
Symphytum	Comfrey
Syringa	Lilac
Tamarix	Tamarisk
Taxus	Yew
Taxus baccata Fastigiata	Irish Yew
Teucrium	Germander
Thuja plicata	Western Red Cedar
Thymus	Thyme
Tilia	Lime
Tilia cordata	Small-Leaved Lime
Tilia platyphyllos	Large-Leaved Lime
Tilia tomentosa	Silver Lime
Trachycarpus fortunei	Chusan Palm
Tsuga heterophylla	Hemlock
Ulex	Gorse

Scientific/Latin	**Common/English**
Viburnum lantana	Wayfaring Tree
Viburnum opulus	Guelder Rose
Viburnum tinus	Lauristinus
Vinca	Periwinkle
Vitis	Vine
Vitis vinifera	Grapevine
Yucca gloriosa	Adam's Needle

Index